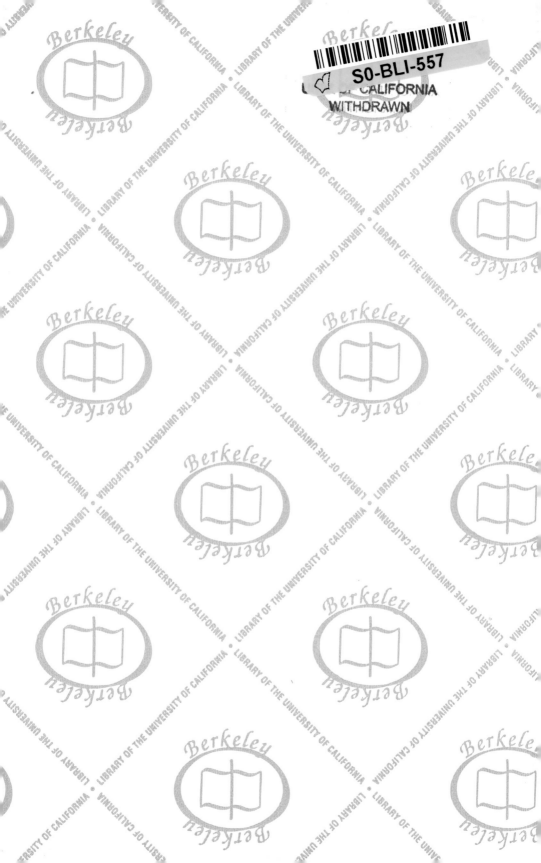

L. Brandsma

Preparative Polar Organometallic Chemistry Volume 2

Experimental Collaboration:
H. Andringa, Y. A. Heus, R. Rikers
L. Tip, H. D. Verkruijsse

Springer-Verlag
Berlin Heidelberg NewYork London
Paris Tokyo HongKong Barcelona

Professor Dr. Lambert Brandsma

Rijksuniversiteit Utrecht
Department of Preparative Organic Chemistry,
Debye Institute
Padualaan 8, NL-3584 CH Utrecht

This work was supported by the Netherlands Technology
Foundation (STW)

MYLAR

ISBN 3-540-52749-4 Springer-Verlag Berlin Heidelberg New York
ISBN 0387-52749-4 Springer-Verlag New York Berlin Heidelberg

Library of Congress Cataloging in Publication Data
Brandsma, L. Preparative polar organometallic chemistry.
Bibliography: v. 1, p.   Includes index.
1. Organometallic chemistry.
I. Verkruijsse, H. D.   II. Title.
QD411.B67   1987   547'.05   86-17896
ISBN 0-387-16916-4 (pbk.: v. 1: U.S.)
ISBN 3-540-52749-4 (Berlin: v. 2)
ISBN 0-387-52749-4 (NewYork: v. 2)

© Springer-Verlag Berlin Heidelberg 1990
Printed in Germany

The useof registered names, trademarks, etc. in this publication does not imply, even
in the absence of a specific statement, that such names are exempt from the relevant
protective laws and regulations and therefore free for general use.

Typesetting: Thomson Press (India) Ltd, New Delhi;
Printing: Mercedes-Druck, Berlin; Bookbinding: Lüderitz & Bauer, Berlin
2151/3020-543210 – Printed on acid-free paper

Preface

This book is a continuation of Vol. 1, which appeared in 1987. It deals with polar organometallic intermediates that can be derived by replacement of an sp^3-proton by an alkali metal atom. A number of these intermediates, among which the acyl-anion equivalents, have shown their usefulness in various organic syntheses published in the last two decades.

The experimental procedures described in this book (which exemplify generally applicable methods) are the result of careful checking of concept-procedures designed by the author: between 1500 and 2000 experiments have been carried out to provide the basis for writing this volume. Just as in Vol. 1, it is described how polar organometallic intermediates can be generated and subsequently be functionalized with a number of "electrophiles". All procedures have been carried out and described on a preparative (usually 0.1 molar) scale, using "preparative" concentrations of starting compounds and reagents (of the order of 0.5 to 0.8 mol/l). Reaction times have been derived from monitoring (in most cases by temperature or colour observation) and hence are a reasonable reflection of reactivities.

The scope of a book like this involves some limitations. Since it would be an impossible task to prepare experimental procedures with all kinds of substrates, we have carried out and described only syntheses with relatively simple and readily available starting compounds. We believe, however, that the reaction conditions described are fairly representative so that they will enable the user of this book to design experimental conditions for many analogous metallation-functionalization reactions on a preparative scale.

The author is indebted to AKZO-Chemicals (Deventer), Andeno (Venlo), Chemetall (Frankfurt am Main, FRG), DSM (Geleen), Diosynth (Oss), and Shell (Amsterdam) for financial support and gifts of chemicals.

Utrecht, November 1990   Lambert Brandsma

Contents

Chapter I
Reactivity of Polar Organometallic Intermediates

1 Introduction

As in the previous volume, only a limited number of procedures for metallations and subsequent functionalizations are given. Although we have tried to choose the procedures as representative as possible, the user of this book may need additional information for designing reaction conditions for a particular synthetic operation. In this respect the present chapter on reactivity of polar organometallic intermediates may be helpful. It is based upon the results of several experiments carried out during the preparation of this volume and other books in this series, and on selected literature data.

2 Alkylation

2.1 Reactivity—A Qualitative Comparison of the Polar Organometallics

In our synthetic investigations we have noticed considerable differences in reactivity of the various types of polar organometallic intermediates towards alkyl halides and epoxides. Exceptionally high alkylation rates were observed in reactions of benzyllithium (or potassium) and allyllithiums (or potassium) with primary alkyl bromides in mixtures of THF and hexane. Under preparative conditions (concentration of reagents ~ 0.5 to 0.8 mol/liter) the characteristic orange colour of benzylalkali solutions disappeared completely within a few seconds upon addition at − 90 °C of a slight excess of alkyl bromide. The allylic intermediates reacted with comparable ease.

It should be pointed out that these alkylations are significantly faster than those of intermediates with comparable basicities having the metal linked to sp^2-carbon, e.g., phenyllithium or -potassium. Many other sp^3-polar organometallic intermediates are more reactive in alkylation reactions than PhLi or PhK, though the differences are not so dramatic as in the cases of benzylic and allylic metal compounds.

Based upon our observations we can make a gross subdivision into the groups A, B, and C for alkylations with alkyl halides and hydroxyalkylations with epoxides in THF-hexane (1:1) mixtures.

A	B	C
Low Reactivity towards alkyl halides	Medium Reactivity towards alkyl halides	High Reactivity towards alkyl halides
$RC\equiv CLi$* enolates Br_2CHLi, Cl_2CHLi Br_3CLi, Cl_3CLi $LiCH(SEt)S(=O)Et$ Cyclopentadienyl-Li	Lithiated nitriles, $RR'C(Li)C\equiv N$ (R and R' = H, alkyl) CH_3SCH_2Li $PhSCH_2Li$ Lithiated aldimines and ketimines Lithiated dihydrooxazines and thiazolines Aryl-Li, Heteroaryl-Li $LiCH_2SOCH_3$ $LiCH(SR)_2$	Aryl-CH_2Li Allyl-Li Heteroaryl-CH_2Li $LiCH_2N=C$

* Not included in this volume.

Within each of the groups there are marked differences. For example, lithium alkynylides, $RC\equiv CLi$, dissolved in mixtures of THF and HMPT (20 vol%) do not react below $-50\,°C$ with ethyl bromide and higher homologues; in the presence of HMPT the carbenoids $LiCCl_3$ and $LiCBr_3$, much weaker bases than acetylides, can be alkylated with good results within one hour at $-100\,°C$. Alkylation of enolates under these extreme conditions proceeds sluggishly.

Depending upon the structural environment, lithium compounds with the system $LiC—C=N—$ show strong differences in reactivity towards alkyl halides. For example, a solution of $LiCH_2CH=N—t$-Bu (or c-hexyl) in THF and hexane reacts very slowly with ethyl bromide and higher homologues at temperatures below $-10\,°C$. Smooth conversions are observed with lithiated 2-methyl-, 3-methyl-, and 4-methylpyridine and lithiated 2-methyloxazines and 2-methylthiazoles, even at temperatures in the range of $-70\,°C$.

Table 1 gives a qualitative comparison of the reactivities of a number of representative polar organometallics towards n-butyl bromide. From this table, 'working' temperature ranges for preparative alkylations of structurally analogous polar organometallics can be derived.

2.2 Scope of the Alkylation Reaction

The "sp^3" organometallic intermediates, which form the subject of this book, give the expected coupling products with primary alkyl bromides, the usual solvent being THF. The use of iodides does not offer special advantages except in alkylations of unstable organometallic compounds such as lithiated 2-methyldihydrooxazines,

Table 1. Reactivity of polar organometallic intermediates towards butyl bromide in 1:1 mixtures of THF and hexane: temperature ranges for butylation with n-C_4H_9Br in preparative experiments

Intermediate[a]	Solution or suspension	Temp. range (°C)	Additional time (min)
$LiCH_2C\equiv N$	susp.	$-50 \rightarrow 0$	30
$LiCH_2N\equiv C$	susp.	$-80 \rightarrow -40$	30
$LiCH_2CH\equiv N-t$-Bu	soln.	$0 \rightarrow 35$	30
$KCH_2CH\equiv N-t$-Bu	soln.	$-30 \rightarrow +10$	20
2- or 4-picolyl-Li	soln. or susp. (resp.)	$-60 \rightarrow -30$	20
$PhCH_2Li$	susp.	$-90 \rightarrow -60$	few sec.
$LiCH(SCH_3)_2$	soln.	$-40 \rightarrow +10$	20
$LiCH(SEt)(SOEt)$	soln.	$+30 \rightarrow +40$	several h
$LiCH_2CH\equiv CHNMe_2$	soln.	$-60 \rightarrow -30$	5
$KCH_2CH\equiv CHO-t$-Bu	susp.	$-60 \rightarrow -30$	5
(structure: oxazoline ring with CH_2Li)	susp.	< -60[b]	1.5 h[b]
$Et_2NCH(Li)C\equiv N$	susp.	$-30 \rightarrow 0$	30
$PhSCH_2Li(TMEDA)$	soln.	$+10 \rightarrow +30$	20
$CH_3SCH_2Li(TMEDA)$	soln.	$+10 \rightarrow +30$	20
$LiCH_2CH\equiv CHSCH_3$	soln.	$-60 \rightarrow -30$	5
$H_2C\equiv C(OCH_3)CH_2K$	susp.	$-80 \rightarrow -30$[c]	10
$LiCH_2CH\equiv CHSiMe_3(TMEDA)$	soln.	$-50 \rightarrow -20$	5
$PhSOCH_2Li$	susp.	$0 \rightarrow +40$	30
(structure: thiazoline ring with CH_2Li)	susp.	$-60 \rightarrow -30$	10
$Alkyl$-$C\equiv CLi$	soln.	> 40	several h
(structure: thiophene with Li)	soln.	$+20 \rightarrow +50$	20
$PhLi$	soln.	$0 \rightarrow +30$	30

[a] Initial concentration 0.8 mol/l; C_4H_9Br (10% excess) is added in one portion at lowest temperature indicated, after which the reaction mixture is kept at the highest temperature of the range for the additional period indicated;
[b] BuBr is added *dropwise* at < -60 °C, since at higher temperature irreversible ring opening occurs;
[c] At higher temperatures CH_3OK is eliminated.

which have to be carried out at temperatures below -60 °C in order to prevent irreversible ring opening, and in alkylations of lithiated S-oxides such as $RC(Li)(SEt)S(\equiv O)Et$, and lithium enolates (which react very sluggishly with alkyl bromides). Chlorides are seldom used.

Formation of 1-alkenes due to β-elimination does not occur with primary halides or is negligible in the reactions of the organometallics of this book. With secondary alkyl halides, which give very low yields with most of the sp- and sp^2-organometallics, the results depend strongly upon the nature of the organometallic intermediate. Alkali-metal compounds that are very reactive towards primary alkyl halides generally give good yields in alkylations with secondary alkyl halides. In the cases of poorly reactive organometallic compounds, e.g., lithium halocarbenoids and enolates, the results with sec-alkyl halides may be unsatisfactory. Yields with

Table 2. Alkylation of polar organometallic compounds with c-hexyl bromide

Intermediate	Reaction conditions	Coupling products yield in %
$NaCH_2C\equiv N$	liq. NH_3, $-33\,^\circ C$	$<5*$
$LiCH_2C\equiv N$	THF-hexane, $0\rightarrow30\,^\circ C$	<10
$KCH_2CH\!=\!N\!-\!c\text{-}C_6H_{11}$	liq. NH_3, $-33\,^\circ C$	10
	liq. NH_3, $-33\,^\circ C$	65
$LiCH(SCH_3)_2$	THF-hexane, $0\rightarrow30\,^\circ C$	<5
$PhCH_2Li$	THF-hexane, $<-30\,^\circ C$	40
$H_2C\!=\!C(CH_3)CH_2K$	THF-hexane, $<-30\,^\circ C$	61
$H_2C\!=\!CHCH(Li)NMe_2$	THF-hexane, $<-30\,^\circ C$	70
$H_2C\!=\!CHCH(K)OCH_3$	THF-hexane, $<-30\,^\circ C$	80
$H_2C\!=\!CH(CH_2K)OCH_3$	THF-hexane, $<-20\,^\circ C$	35
$H_2C\!=\!CHCH(Li)SCH_3$	THF-hexane, $<-30\,^\circ C$	78
	liq. NH_3, $-33\,^\circ C$	82

* Mainly cyclohexene is formed.

cyclopentyl bromide are usually higher than 50%, in the case of cyclohexyl bromide they vary strongly with the organometallic intermediate (see Table 2).

Some reactions with benzyl chloride give low yields of impure alkylation products due to formation of stilbene. In these cases it is advisable to repeat the synthesis with the *bromide*.

With 1,2-dihaloethane the results of coupling reactions are generally poor due to extensive β-elimination or attack of the 'anion' on halogen.

Substitutions with bromoacetaldehyde diethylacetal, $BrCH_2CH(OEt)_2$, proceed much less smoothly than with the normal primary alkyl bromides. Furthermore, these reactions are often accompanied by elimination of HBr and HOEt, which reduces the yields of the desired products.

Tertiary alkyl halides usually give only a few percent of the *tert*-butyl derivative upon interaction with an alkali-organic compound. The incidental couplings with satisfactory yields are likely to be the result of a radical-like mechanism.

2.3 Dialkylation

During some alkylation reactions, undesired introduction of two (or three) alkyl groups occurs, while the mono-alkyl derivative is obtained in reduced yields. The following example is illustrative (compare Chap. VIII):

$$RCH_2C\equiv N \xrightarrow[\text{THF-hexane}]{\text{LDA}} RCH(Li)C\equiv N \xrightarrow{R'Br} RCH(R')C\equiv N \rightleftharpoons$$
$$(I) \qquad\qquad\qquad\qquad I$$

$$RC(Li)(R')C\equiv N \xrightarrow{R'Br} RC(R')_2C\equiv N$$
$$(+\, RCH_2C\equiv N)$$

This situation can occur when the mono-alkyl product is sufficiently acidic to equilibrate with the initial intermediate (I). Reaction of ketone and nitrile anions with alkyl halides often gives considerable amounts of dialkyl derivatives. Addition of the metallated ketone or nitrile to an excess of alkyl halide ('inversed-order' addition) gives no significant improvement. Dialkylation is much less serious in the cases of strongly basic organometallics, which usually react vigorously with alkyl halides. The complication of dialkylation may arise also, when the organometallic intermediate is sparingly soluble: reaction of the suspension of dimsyllithium, $CH_3S(=O)CH_2Li$, in THF-hexane with alkyl- or allyl bromide proceeds smoothly, but the product is a mixture of comparable amounts of $RCH_2S(=O)CH_3$ and $RCH_2S(=O)CH_2R$.

2.4 Remarks on the Reaction Conditions of Alkylations

a) Solvent

The usual solvent for alkylations of the 'sp^3' polar organometallics is THF (in admixture with hexane). The alkylations with bromides and iodides usually proceed at a sufficiently high rate at temperatures below $0\,°C$, some require temperatures in the region of $0°$ to $30\,°C$. Some chemists use HMPT as a co-solvent, but in fact this can be omitted in most cases; it is indispensable, however, in alkylations of the thermolabile carbenoids (e.g., $LiCBr_3$), which have to proceed within a relatively short period at very low temperatures. Also, reactions with $BrCH_2CH(OR)_2$, which usually proceed sluggishly, are accelerated considerably by adding a certain volume of HMPT.

The highly reactive intermediates of group C (see p. 2) react smoothly in Et_2O or sometimes even in hexane, provided that they are in a dissolved state in these solvents.

Liquid ammonia is a very good solvent for alkylations and β-hydroxyalkylations (with epoxides) and its use deserves recommendation for alkalimetal compounds that react sluggishly in THF. Alkylation of a suspension of cyclopentadienyllithium (CpLi) in THF with primary alkyl halides is extremely slow. The solution of CpLi in liquid ammonia reacts smoothly with primary alkyl bromides including the slightly soluble higher homologues, e.g., n-$C_6H_{13}Br$. The reaction with this halide is complete within fifteen minutes and gives the hexyl derivative in a reasonable yield. Another illustrative example is the alkylation of the lithiated S-oxide $LiCH(SC_2H_5)(SOC_2H_5)$ which has a good solubility in THF as well as in liquid ammonia. Whereas completion of the reaction of this intermediate in THF with butyl bromide at $+ 20$ to $30\,°C$ takes several hours, the alkylation in boiling (b.p. $\sim - 33\,°C$) ammonia proceeds almost instantaneously. In Chap. X some efficient procedures for the alkylation of enolates in liquid ammonia are described.

Many of the organometallic compounds in this book are stable (i.e. are not ammonolysed) in liquid ammonia. They can be prepared simply by adding the substrate to an equivalent amount (or slight excess) of an in situ prepared alkali amide in this solvent.

$$RH + MNH_2 \rightleftharpoons RM + NH_3$$

The deprotonations are usually very fast in the polar liquid ammonia (within a few minutes). The success of an alkylation in liquid ammonia depends *inter alia* upon the position of the deprotonation equilibrium. If the acidities (pK) of the substrate and NH_3 (pK $\sim$ 34) are comparable, the equilibrium may be on the left side because $LiNH_2$ and $NaNH_2$ are poorly soluble in ammonia. However, when RM is a strong C-nucleophile and soluble in ammonia the alkylation may be successful since the reaction with the slightly soluble $LiNH_2$ or $NaNH_2$ is relatively slow. A typical example is the alkylation of $H_2C(SC_2H_5)_2$ with ethyl bromide and higher homologues in the presence of sodamide in liquid ammonia. Although the 'ionization' of the mercaptal with $NaNH_2$ is incomplete, yields of $RCH(SC_2H_5)_2$ may be good to excellent, especially when an excess of $NaNH_2$ (and a corresponding excess of alkyl bromide) is used.

b) Counterion

It is usually observed that reactions of organo-potassium compounds with alkyl halides in THF-hexane mixtures proceed within temperature ranges that are by some 40 °C lower than in the cases of the corresponding lithium compounds. This knowledge may be applied if, in view of limited stability of the intermediate or for some other reason, the alkylation has to be carried out at the lowest possible temperature. Lithium can be replaced easily by potassium by adding an equivalent amount of t-BuOK to the solution of the lithium compound in THF. Direct generation of the organo-potassium intermediate is in many cases possible by metallating the starting compound with a mixture of BuLi and t-BuOK at low temperatures.

3 Hydroxyalkylation with Epoxides

In their behaviour towards polar organometallic compounds epoxides and alkyl halides show a rather strong resemblance. In both cases the reaction is facilitated by increasing the polarity of the solvent or the size of the counterion. Orders of reactivity of the various polar organometallic intermediates towards alkyl halides and towards epoxides are roughly similar. 'Preparative' conversions of organo-alkali compounds with epoxides in THF proceed within the same (or slightly higher) temperature range as those with primary alkyl bromides, RBr (R $\geqslant C_2H_5$) (see Table 1). Under less polar conditions (e.g., in Et_2O), however, the hydroxyalkylation is assisted by coordination of the lithium ion with the epoxide-oxygen. This may lead to the situation that in Et_2O the hydroxyalkylation with oxirane is faster than the alkylation with a homologue of methyl bromide, whereas in a more polar medium both reactions may proceed at comparable rates, or the reaction with the alkyl bromide may be even faster.

Under the usual conditions (THF or liquid ammonia as solvent), mono-substituted epoxides are attacked on the methylene carbon atom. The stereochemistry of the ring opening of vicinally di-substituted epoxides (e.g., epoxycyclohexane) is in accordance with an S_N2-mechanism.

4 Hydroxyalkylation with Carbonyl Compounds

Reactions of organolithium compounds with aldehydes and ketones in THF or Et_2O are usually extremely fast, even at temperatures in the region of $-80°$ to $-100°C$. Reaction times for $> 95\%$ conversion at concentrations 0.5 to 0.8 mol/l are of the order of a few seconds to a few minutes at $\sim -70°C$, though slightly soluble species may react somewhat less fast. Yields of the expected carbinols are generally good to excellent. Interaction between organosodium or -potassium compounds and enolizable carbonyl compounds very often gives rise to extensive enolate formation. Good yields of carbinols from reactions in liquid ammonia or in THF-HMPT mixtures are rather exceptional (see, for example, Ref. [1]).

For the introduction of a hydroxymethyl group our procedure with dry polymeric formaldehyde (paraform) in many cases works satisfactorily [2]. We have noticed that, irrespective of the nature of the organometallic species (mostly Li as counterion), the reaction in Et_2O or THF does not start below 15 °C. This might be explained by assuming that at room temperature a sufficiently high rate of dissociation into the monomer is attained. Liquid ammonia is generally unsuitable as a solvent for the hydroxymethylation with paraformaldehyde.

Literature

1. Brandsma L (1988) Preparative Acetylenic Chemistry, Revised Edition, Elsevier
2. Schaap A, Brandsma L, Arens JF (1965) Recl Trav Chim Pays-Bas 84:1200

5 Formylation with Dimethylformamide

Introduction of an aldehyde function may be accomplished by adding DMF to a solution of the organolithium compound in Et_2O or THF and subsequently hydrolysing the resulting solution with dilute aqueous acid:

$$RLi + HC(=O)NMe_2 \longrightarrow RCH(OLi)NMe_2 \xrightarrow{H^+, H_2O} RCH=O$$

With strongly basic species such as $(CH_3S)_2CHLi$ and $Aryl-CH_2Li$, the reaction with DMF is almost instantaneous at temperatures between -60 and $-80°C$ (in the case of $Aryl-CH_2Li$ visible by the immediate discharge of the intensive colour). For reasons given below it seems essential to prevent the aqueous mixture from becoming basic (pH > 7) during the hydrolysing operation. The use of a (slight) excess of acid and efficient stirring are therefore necessary. In the presence of traces of base, aldehydes $R'CH_2CH{=}O$ in which $R' = Aryl$ or in which R' contains a double bond adjacent to the CH_2-group, may readily undergo aldol-condensation. If R' contains a tertiary amine function (e.g., if $R' = $ pyridyl) the aldehyde dissolves in the acid medium and the base necessary to liberate the product is likely to destroy it. In connection with these practical problems, the applicability of this formylation method is more limited than in the cases of sp- and sp^2-organolithium compounds.

6 Carboxylation

The usual protocol for carboxylation of an organometallic intermediate involves pouring the solution onto dry solid carbon dioxide (preferably powdered), in a wide-necked conical flask or beaker. It is advised to cover the carbon dioxide with an organic solvent, e.g., Et_2O. It is not clear, however, whether the CO_2 is crushed 'dry ice' or the powder, obtained from a fire extinguisher. In the first case one may seriously wonder if no moisture has condensed in the 'dry ice'. If relatively small volumes of solution have to be carboxylated, the presence of frozen H_2O in the 'dry ice' probably will not reduce the yield of the carboxylic acid, since quenching of the organometallic compound by the gaseous CO_2 evolved will be complete before the frozen water has dissolved into the organic solvent and protonated the intermediate. In our opinion the reaction with CO_2 can be carried out more sophisticatedly by slowly introducing the solution of the organoalkali compound by means of a syringe into a vigorously stirred solution of a large excess of CO_2 in THF, kept below $-80\,°C$. By keeping the concentration of CO_2 very high, the following conversion, in which maximally 50% of the original lithium compound is consumed, can be prevented:

$$RR'CHC({=}O)OLi \xrightarrow{\ RR'CHLi\ } RR'C{=}C(OLi)_2$$

7 Reaction of Organoalkali Compounds with Carbon Disulfide

The reaction of 'sp^3' organoalkali compounds with carbon disulfide usually does not stop at the stage of the initial adduct, but is followed by an extremely fast conversion of the dithiocarboxylate into a geminal ene-dithiolate:

$$RR'CHM \xrightarrow{\ CS_2\ } RR'CHC({=}S)SM \xrightarrow{\ RR'CHM\ } RR'C{=}C(SM)_2$$

$$(M = Li,\ Na,\ K)$$

This further reaction is difficult to suppress, even when the organometallic compound is added to a solution of CS_2 in the organic solvent.

8 Addition of Organoalkali Compounds to Isocyanates and Isothiocyanates

Addition of 'sp^3' organoalkali compounds to isocyanates and isothiocyanates have been reported incidentally.

$$RR'CHM + R''N{=}C{=}X \longrightarrow RR'CHC({=}NR'')X^-$$

$$\xrightarrow{\text{H}_2\text{O}} RR'CHC({=}X)NHR''$$

$$(X = O \text{ or } S)$$

With strongly basic organometallics (e.g., $(CH_3S)_2CHLi$) the additions are extremely fast, even at very low temperatures. Yields are mostly excellent.

9 Sulfenylation

A synthetically useful reaction is the formation of sulfides from organoalkali compounds and disulfides:

$$RR'CHM + R''SSR'' \longrightarrow RR'CHSR''$$

For the sulfenylation of the less strongly basic ketone-enolates the more reactive thiosulfonates $R''SSO_2R''$, thiocyanates $R''SC{\equiv}N$ or sulfenyl halides $R''SCl$ (or Br), may be used instead of dialkyl disulfides. In many cases, however, the reaction with the more easily obtainable disulfides is sufficiently fast. Di-*tert*-butyl disulfide, *t*-BuSS-*t*-Bu reacts very sluggishly, even with the most strongly basic organometallics.

If the normal addition procedure is applied, the initial product may be deprotonated by the original organometallic compound and the newly generated intermediate will react with the disulfide:

$$RR'CHSR'' \xrightarrow{RR'CHM} RR'C(M)SR'' \xrightarrow{R''SSR''} RR'C(SR'')_2$$

In some cases this further reaction can be suppressed by adding the sulfenylating agent (preferably in excess) in one portion, but a more successful procedure involves addition of RR'CHM to a solution of a (100%) excess of R''SSR''. If the disulfide is non-volatile (PhSSPh!) and hence the excess is difficult to remove from the product mixture, this procedure is not attractive. Especially in the cases of metallated nitriles $RCH_2C{\equiv}N$ and ketone enolates the further sulfenylation is a problem which has not yet been satisfactorily solved.

10 Trimethylsilylation

Trimethylchlorosilane is routinely used to introduce trimethylsilyl groups by its reaction with organometallic intermediates. In some reactions, however, complications may arise when an excess of this reagent is employed. In conversions of organolithium compounds having the structural system $Li-C-C=N$ we obtained significantly lower yields of the desired products when an excess of trimethylchlorosilane was used. This may be explained by assuming that the hydrochloric acid formed from the excess of silylating agent during aqueous work-up, protonates nitrogen, after which water (or ^-OH) can cause desilylation as visualized in the following example:

$$Me_3SiCH_2CH=N-t\text{-}Bu \xrightarrow{H^+} Me_3Si-CH_2-CH=\overset{\overset{\displaystyle H}{|}}{N^+}-t\text{-}Bu \longrightarrow$$

$$\downarrow H_2O$$

$$\longrightarrow H_2C=CH-NH-t\text{-}Bu \longrightarrow H_3CCH=N-t\text{-}Bu$$

A similar reaction occurs in the cases of trimethylsilylation of 2- or 4-picolyllithium or lutidyllithium, 2-lithiomethyl thiazoline, and 2-lithiomethyl dihydrooxazines. This undesired process can be avoided if, prior to aqueous work-up, the greater part of the solvent is removed under reduced pressure, so that also the excess of Me_3SiCl is swept along with the solvent. Instead of applying this procedure, the excess of Me_3SiCl can be destroyed by adding (at room temperature) a corresponding amount of diethylamine. Diisopropylamine is not suitable since its reaction with Me_3SiCl is too slow.

11 Reactions of Organometallic Compounds with Chloroformates and Dimethylcarbamoyl Chloride

Acetylides $RC\equiv CM$ and several metallated 'sp^2' compounds can be successfully functionalized with chloroformic esters $ClCOOR$, and the carbamoyl chloride $ClC(=O)NMe_2$. In order to avoid subsequent reactions of the products with the organometallic compound, the latter is added to a large (usually 50 to 100 mol%) excess of the reagent. The methylene or methyne protons in the initial product formed from an 'sp^3' organometallic compound with $ClCOOR$ or $ClC(=O)NMe_2$ are more acidic than those in the starting compound, e.g.:

$$(CH_3S)_2CH_2 \xrightarrow{BuLi} (CH_3S)_2CHLi \xrightarrow{ClCOOR} (CH_3S)_2CH-COOR \xrightarrow{1}$$

$$\text{(I)}$$

$$(CH_3S)_2C=C(OLi)OR \xrightarrow{ClCOOR} (CH_3S)_2C(COOR)_2$$

As a consequence, the initial product will be rapidly metallated by I, after which the metallation product may react with the chloroformic ester. These subsequent processes, together with attack of I on the COOR-group in the initial product, may be responsible for the very low yield of $(CH_3S)_2CHCOOC_2H_5$, even when the lithiated S,S-acetal is added with strong cooling to a 100% excess of the chloroformate. Other strongly basic organometallics, e.g., Aryl-CH_2Li, also give poor yields in reactions with ClCOOR. The use of a fifty-fold excess of chloroformate will undoubtedly lead to improved results, but is not very attractive for preparative-scale syntheses.

With dimethylcarbamoyl chloride, the same lithiated S,S-acetal gave only 45% yield of $(CH_3S)_2CHC(=O)NMe_2$ in a similar inversed-order addition procedure. In this case, the product arising from transmetallation, $(CH_3S)_2C=C(NMe_2)OLi$, will probably not react very easily with $ClC(=O)NMe_2$.

In the reactions of metallated methylpyridines, imines, thiazolines, and dihydro-oxazines, quaternization reactions with the acylating reagent may form an additional problem.

12 Reactions of Organoalkali Compounds with Halogenating Agents

Many sp- and sp^2-polar organometallic compounds can be successfully converted into sp- and sp^2-halogen derivatives by reaction with free halogen or a suitable halogenating agent:

$$-C\equiv C-M + X_2 \longrightarrow -C\equiv C-X + MX$$

$$-C=C-M + X_2 \longrightarrow -C=C-X + MX$$

Attempts to halogenate sp^3-alkalimetal compounds in many cases lead to the formation of 'dimeric' products as a consequence of a very fast subsequent reaction of the organometallic compound with the halogen derivative initially formed. Effective suppression of this reaction is only possible in conversions of less strongly basic organometallic compounds (e.g., enolates) with iodine: the alkali-metal compound has to be added to the strongly cooled solution of iodine in Et_2O or THF.

13 Conjugate Additions

The addition of (strongly basic) organoalkali-metal compounds to conjugated diene systems, in general, has little synthetic value, as the reaction usually results in the formation of polymers [1]. Additions of alkyllithium to vinylic silanes or sulfides, $H_2C=CH-X$ (X = Me$_3$Si or SPh) [2], ketene-S,S-acetals $H_2C=C(SR)_2$ [3] and to conjugated systems such as $>C=CH-CH=C(SR)_2$ [4] and $H_2C=CH-CH=CHSR$ (1,4-additions) [5] are generally successful. The orientation in these reactions is such that the alkyl group attaches to the β-carbon or S-carbon atom, respectively. Enyne compounds $H_2C=CHC\equiv CR$ (R = alkyl, SR, SiMe$_3$, CH$_2$NR$_2$) undergo 1,4-addition of strongly basic organoalkali derivatives R'M whereby (after protonation) allenes, $R'CH_2CH=C=CHR$, are obtained [6].

A very useful reaction in organic synthesis is the so-called Michael addition of an organic compound RH to an α,β-unsaturated system $C=C-X$, in which X may represent an electron-withdrawing group such as $C=O$, COOR, $C\equiv N$, NO$_2$, S($=$O)R. The addition, in which R attaches to the β-carbon atom to give $R-C-CH-X$, occurs in a protic medium with sufficiently acidic compounds, e.g., nitroalkanes, alcohols, thiols, malonic esters and requires a basic catalyst.

The basicities of a number of synthetically important anions (e.g., $\underset{\cdot}{C}H(SR)_2$, $R\underset{\cdot}{C}H-C\equiv CN$, $R\underset{\cdot}{C}H-CH=N-t$-Bu) are too high, however, to generate them in a sufficiently high concentration using a base in a catalytic amount. Under more forcing conditions, with a base like t-BuOK in DMSO as a catalyst, the Michael-acceptor is likely to undergo undesired reactions. It is therefore necessary to generate the anion in a separate step by deprotonating the donor RH with a stochiometrical amount of a strong base under aprotic conditions, after which the Michael-acceptor is added. The final step involves the addition of a sufficiently acidic proton donor, usually water, e.g.:

$$RH + BM \longrightarrow R^-M^+ + HB$$

$$\overset{}{\underset{}{>}}C=C-C=O + R^-M^+ \longrightarrow R-C-C=C-OM$$

$$R-C-C=C-OM \xrightarrow{\; H^+(H_2O)\;} RC-CH-C=O$$

There are a number of additional possibilities for interaction between the organometallic intermediate RM and the Michael-acceptor, the most general one being addition across the $C=O$ bond, usually referred to as 1,2-addition. The Michael-acceptor may also be deprotonated by RM, if acidic protons are present, e.g.:

Reaction of an allylic metal compound like $H_2\overset{\gamma}{C}=CH-\overset{\alpha}{C}H(Li)SPh$ with an enone $\overset{4}{C}=\overset{3}{C}-\overset{2}{C}=\overset{1}{O}$ may give four addition products, since bonds can be formed between C-α and C-2 and C-4 as well as between C-γ and C-2 and C-4 [7]. With metallated thiocarbonyl compounds, $(C=C-S^- \leftrightarrow {}^-C-C=S)M^+$, S-1,4- as well as C-1,4-addition has been observed [8].

Several papers in the literature of the last fifteen years deal with the problem of combining organometallic intermediates and α,β-unsaturated systems in a 1,4-fashion. With organocopper compounds this reaction is, in general, successful, but with polar organometallic intermediates RM the results (1,2- or 1,4) are very strongly dependent upon the structures of the group R and the acceptor. Substituents can have a profound influence upon the ratio of 1,4- and 1,2-addition.

Increase of the polarity of the solvent can have opposite effects with regard to the regiochemistry with two different polar organometallic intermediates. Some examples are given in the review listed as Ref. [9], others can be found in "Annual Reports in Organic Synthesis", under the heading "Conjugate Addition".

Although no generally applicable experimental rules can be given for controlling the regiochemistry of reactions with polar organometallic derivatives and α,β-unsaturated compounds, some trends and regularities can be observed:

1. 1,4-Additions of sp- and sp^2-organometallics have not been reported so far.
2. Organometallic compounds (especially the less strongly basic ones) in which the anion is stabilized by electron-withdrawing, thioether or trialkylsilyl groups ($C\equiv N$, $COOR$, NO_2, SO_2R, SR, $SiMe_3$, $P(=O)R$, etc.) are often good partners for 1,4-additions.
3. α,β-Unsaturated esters, $>C=C-COOR$, and amides, $>C=C-C(=O)-NR_2$, (or thioamides) are much better Michael-acceptors than α,β-unsaturated aldehydes or-ketones: their $C=O$ group is less prone to undergo 1,2-addition.
4. In some reactions the 1,2-addition is kinetically controlled and reversible, so that longer reaction times or higher temperatures may result in the formation of an increased amount of 1,4-adduct. In some cases a similar effect can be attained by performing the addition in the presence of a certain amount of hexamethylphosphoric triamide (HMPT). This solvent effect may be the consequence of a reversal of the 1,2-addition, but there are also cases in which HMPT promotes a *kinetically* controlled 1,4-addition [10].

Literature

1. Wakefield BJ (1974) The chemistry of organolithium compounds, Pergamon Press; (1988) Organolithium methods, Academic Press
2. Ager DJ (1981) Tetrahedron Lett 587
3. Carlson RM, Helquist PM (1969) Tetrahedron Lett 173; Gröbel BT, Seebach D (1977) Synthesis 392
4. Gröbel BT, Seebach D (1977) Synthesis 394
5. Brandsma L, Everhardus RH (unpublished observations)
6. Wakefield BJ (1982) In: Comprehensive organometallic chemistry, vol 7, Pergamon Press
7. Binns MR, Haynes RK, Houston TL, Jackson WR (1980) Tetrahedron Lett 575
8. Berrada S, Metzner P (1986) Bull Soc Chim France 817; see also Kpegba K, Metzner P, Rakatonirina R (1986) Tetrhedron Lett 1505
9. Albright JD (1983) Tetrahedron 39:3211
10. Wartski L, EL Bouz M, Seyden-Penne J (1979) Tetrahedron Lett 1543; Sauvetre R, Roux-Schmitt MC, Seyden-Penne (1978) Tetrahedron 34:2135; Roux-Schmitt MC, Wartski L, Seyden-Penne J (1980) J Chem Research (S) 346; Ostrowski PC, Kane VV (1977) Tetrahedron Lett 3549

Information regarding the handling of organolithium compounds on laboratory and industrial scale is supplied by Chemetall GmbH, Reuterweg 14, D6000 Frankfurt am Main, FRG

Chapter II
Metallation of Aromatic and Olefinic Hydrocarbons

1 Introduction

Unactivated olefins and aromatic compounds, ArCHRR', are among the least 'acidic' compounds. Their complete conversion into the corresponding alkali-metal compounds requires the use of butyllithium or reagents derived therefrom. In many cases an (preparatively) acceptable rate of metallation can be attained only when using a large excess of olefin or aromatic compound. Considerable progress in this field was made by the development of the super-basic reagent BuLi · t-BuOK. After the first paper on this reagent by Lochmann et al. [1] Schlosser and co-workers published a number of successful metallations in the aromatic and olefinic series [2]. Further possibilities for metallation arise with the variant BuLi·t-BuOK·TMEDA-hexane developed by Brandsma Schleyer, et al. [3]. In Tables 3 and 4 preparative metallation conditions for a number of aromatic and olefinic compounds are summarized. Most of these metallations are described in detail in the experimental section.

Literature

1. Lochmann L, Pospisil J, Lim D (1966) Tetrahedron Lett 257
2. Schlosser M, Hartmann J (1973) Angew. Chemie 85:544; (1973) Int Ed (Engl) 12:508 and subsequent papers by Schlosser et al.; see also the review on superbases Schlosser M (1988) Pure and Applied Chem 60:1627
3. Brandsma L, Verkruijsse HD, Schade C, von R Schleyer P (1986) J Chem Soc, Chem Comm 260

2 Metallation of Alkylbenzenes and Alkylnaphthalenes

Lithiation of toluene, o-, m-, or p-xylenes and mesitylene can be readily achieved by treating the hydrocarbons (in large excess) with a 1:1 molar mixture of BuLi and TMEDA in hexane at slightly elevated temperatures or under reflux. Whereas the metallation of the xylenes and mesitylene by this procedure is highly regioselective, toluene gives 10 to 15% or ring-lithiated products [1, compare 2] in addition to benzyllithium. In the cases of 1- and 2-methylnaphthalene, methyl- and ring-lithiation occur to a comparable extent. Specific α-metallation of primary alkyl-benzenes and the methylnaphthalenes can be achieved with the Lochmann–Schlosser reagent [3–6] or our variant BuLi · t-BuOK · TMEDA (1:1:1) in hexane

[7]. Treatment of ethylbenzene with these reagents initially gives much ring-metallated compound (presumably *meta* and *para*), but the formation of the α-potassiated derivative, PhCH(K)CH$_3$ is favoured thermodynamically. Its formation from the ring-metallated intermediate presumably occurs by a process of proton-donation (ethylbenzene being the proton donor) and -abstraction [1]:

The 1:1 molar mixture of BuLi and *t*-BuOK has been used by Schlosser and Hartmann for the preparation of 2-naphthylmethylpotassium from 2-methyl-naphthalene [5].

n-Amylsodium and n-amylpotassium [8, 9], in combination with TMEDA, have shown to be extremely effective reagents for the α-metallation of aromatic hydrocarbons. Even isopropylbenzene, which is very difficult to metallate at the tertiary carbon atom with the Lochmann–Schlosser reagent, is reported to give exclusively PhC(K)(CH$_3$)$_2$. The formation of this compound is preceded by the kinetically controlled formation of *m*- and *p*-(ring)metallated sodium and potassium compounds. It may be concluded that the n-amylsodium and -potassium reagents are stronger bases (in a kinetic sense) than the 1:1 molar mixture of n-BuLi and *t*-BuOK. Unfortunately the successful preparation of n-amylsodium and -potassium from amylchloride and alkali-metal dispersion is not easy to realize. Another very powerful reagent capable of converting cumene into PhC(K)(CH$_3$)$_2$, is trimethylsilylmethylpotassium [10]. It has to be prepared from the mercury derivative [Me$_3$SiCH$_2$]$_2$Hg, and therefore is of limited preparative use.

In a recent article [11] Ahlbrecht et al. described in metallation of toluene and some other aromatic hydrocarbons with potassium dialkylamides, KNR$_2$, (e.g., R = *i*-C$_3$H$_7$, *c*-C$_6$H$_{11}$, or R$_2$ = 2,2,6,6-tetramethylpiperidyl) in THF. It was concluded from NMR-spectra that toluene is completely metallated by KTMP and for more than 60% by KDA. Reaction of the red solution with alkyl chlorides and oxirane gave the expected products PhCH$_2$R and PhCH$_2$CH$_2$CH$_2$OH in good yields. We could confirm the latter results, but obtained moderate and very low yields from the reactions of toluene with KTMP and KDA, respectively, and subsequent quenching with trimethylchlorosilane and dimethyldisulfide.

Literature

1. Unpublished observations in the author's laboratory
2. Eberhardt GG, Butte WA (1964) J Org Chem 29:2928
3. Lochmann L, Pospísil J, Lim D (1966) Tetrahedron Lett 257
4. Schlosser M (1967) J Org Chem 8:9
5. Schlosser M, Hartmann J (1973) Angew Chemie 85:544; (1973) Int Ed (Engl) 12:508
6. Schlosser M, (1973) Polare Organometalle, Springer, Berlin Heidelberg New York
7. Brandsma L, Verkruijsse HD, Schade C, von R Schleyer P (1986) J Chem Soc, Chem Comm 260

8. Crimmins TF, Chan CM (1976) J Org Chem 41:1870
9. Trimitsis GB, Tuncay A, Beyer RD, Ketterman KJ (1973) J Org Chem 38:1491
10. Hartmann J, Schlosser M (1976) Helv Chim Acta 59:453
11. Ahlbrecht H, Schneider G (1986) Tetrahedron 42:4729

3 Dimetallation of Aromatic Compounds

Trimitsis and co-workers [1] reported that the formation of α,α'-disodium compounds of 1,2-, 1,3-, 1,6-, and 1,8-dimethylnaphthalene and of o- and m-xylene proceeds with quantitative yields, when the hydrocarbons are treated with butylsodium in the presence of TMEDA. 1,4-Dimethylnaphthalene and p-xylene could not be converted into the disodium compounds. Our attempts [2] to dimetallate o- and m-xylene with the much more easily available 1:1:1 molar mixture (2 equivalents) of BuLi, t-BuONa, and TMEDA in hexane gave mainly the mono-metallated derivatives. Using a 30 mol% excess of the more strongly basic reagent BuLi · t-BuOK · TMEDA we obtained a 80:20 mixture of α,α'-dimetallated and mono-α-metallated compound from m-xylene.

The paper of Bates and Ogle [3] mentions the successful formation of the p-xylene dianion from p-xylene and the Lochmann–Schlosser base, BuLi · t-BuOK,

Table 3. Representative α-metallations of aromatic hydrocarbons*

Substrate (mol% excess)	Base-Solvent System	Temperatures (°C) (Reaction time in min)	Remarks
Ph-CH$_3$ (300)	BuLi · t-BuOK · TMEDA, hexane	$-20(30) \rightarrow +10(60)$**	
Ph-CH$_3$ (200)	BuLi · t-BuOK, THF-hexane	$-100(1) \rightarrow 0(0)$	a
Ph-CH$_3$ (300)	BuLi · t-BuOK, hexane	$+20(15) \rightarrow 40(45)$	
Ph-CH$_2$CH$_3$ (300)	BuLi · t-BuOK · TMEDA, hexane	$-20(45) \rightarrow +15(90)$	b
Naphthyl-1-CH$_3$ or 2-CH$_3$	BuLi · t-BuOK · TMEDA, hexane	$-20(45) \rightarrow +10(10)$	c
Xylene (o-, m-, or p-) (200)	BuLi · TMEDA, hexane	$+30(1) \rightarrow 70(30)$	
Xylene (m-) (10 to 50)	BuLi · t-BuOK · TMEDA, hexane	$-20(30) \rightarrow +10(15)$	
Xylene (m-) (0)	BuLi · t-BuOK · TMEDA, hexane (dimetallation)	$-25(120) \rightarrow +10(15)$	d
Mesitylene (100)	BuLi · TMEDA hexane	$30(1) \rightarrow 70(30)$	
Ph$_3$CH	NaNH$_2$ or KNH$_2$, liq. NH$_3$	$-33(0)$	
Ph$_2$CH$_2$	KNH$_2$, liq. NH$_3$	$-33(0)$	

* For experimental details see experimental section;
** $\rightarrow$ means allowing the temperature to rise without cooling bath;
a Ratio THF/hexane (v/v) $\sim$ 5; lower ratios gave reduced yields;
b Initially much meta- and para-metallated ethylbenzene is present;
c $\sim 20\%$ excess of the basic reagent was used;
d For this dimetallation ~ 30 mol% excess of the basic reagent was used.

in hexane, which stands in sharp contrast to the failure reported by Trimitsis et al. [1].

Klein and co-workers [4] needed a large excess of BuLi·TMEDA for the α,α′-dilithiation of *m*-xylene. In their experiments they obtained comparable amounts of mono-(α)-metallated product. Attempts to lithiate both CH_3-groups in *p*-xylene failed: the main products were the mono-lithiated and geminally di-lithiated derivatives. Treatment of *o*-xylene and mesitylene with an excess of BuLi·TMEDA was reported to give product mixtures, containing mono- and geminally di-lithiated compounds in addition to the desired α,α′- or α,α′,α″-polylithiated derivatives.

Summarizing, one may conclude that there is no simple and preparatively attractive method available for the generation of α,α′-di-alkalimetal compounds from the isomeric xylenes and mesitylene.

Literature

1. Trimitsis GB, Tuncay A, Beyer RD, Ketterman KJ (1973) J Org Chem 38: 1491
2. Unpublished results from the author's laboratory
3. Bates RB, Ogle CA, (1982) J Org Chem 47:3949
4. Klein J, Medlik A, Meyer AY (1976) Tetrahedron 32:51

4 Metallation of Olefinic Compounds

Some representative structures of olefinic hydrocarbons are given below. They are placed in order of increasing kinetic acidity. This order is based on literature data, on our experiments and on general knowledge on inductive stabilization or destabilization of negative charge. The carbon atoms from which preferential abstraction of a proton occurs are underlined, pertinent literature is indicated in brackets under the formula.

[1]

Alkyl-$\underline{C}H_2$-CH=CH-CH_2-Alkyl
[4]

[2,3]

$\underline{C}H_3CH=CHCH_2$-Alkyl
[4]

$H_2C=CH\underline{C}H_2$-Alkyl
[4]

$(\underline{C}H_3)_2C=CHCH_3$
[2]

$(CH_3)_2C=CHCH_2CH_2C(\underline{C}H_3)=CH_2$
[6]

[5]

$\underline{C}H_3CH=CHCH_3$
[4,8]

$H_2C=CH\underline{C}H_3$
[7]

$H_2C=C(\underline{C}H_3)_2$
[7,15]

[14]

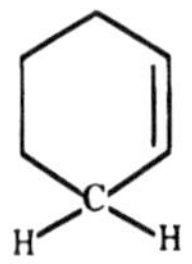

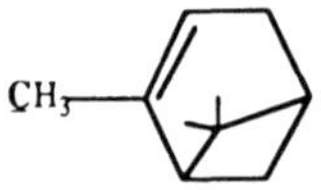

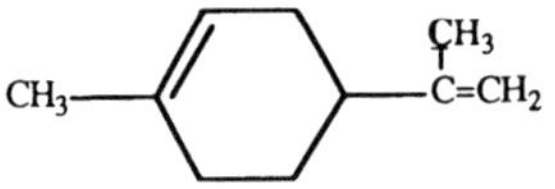

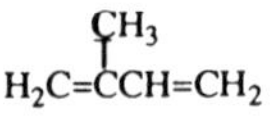

$CH_3C=CHCH=CH_2$ $H_2C=CHCH_2CH=CH_2$

[11,16]   [10,12,13]

$H_2C=CHCH_2Ph$ $CH_3(CH=CH)_2CH=CH_2$

[12]

[17,18]

Using superbasic reagents such as BuLi·t-BuOK and Me_3SiCH_2K, olefins with a poor kinetic acidity can be metallated relatively easily [1–4]. BuLi·TMEDA is a less efficient deprotonating reagent, though good results have been reported with the metallation of limonene [5], propene, and isobutene [7].

Conjugated or 'skipped' dienes and trienes undergo ionization with alkali amides in liquid ammonia. In the cases of allylbenzene, heptatriene, indene, and cyclopentadiene, the equilibrium $RH + MNH_2 \rightleftarrows RM + NH_3$ is probably completely on the right side (with M = Na or K), but the thermodynamic acidities of cyclohexadiene and pentadiene are probably very close to that of ammonia, so that metallation with alkali amides (even potassium amide) may be incomplete.

Isoprene undergoes 'anionic' polymerization upon interaction with BuLi or variants. Although potassium diisopropylamide (KDA) and potassium tetramethylpiperidide (KTMP) are not sufficiently strong bases to cause complete metallation of isoprene, subsequent reactions of the reaction mixtures with alkyl halides and oxirane give the expected derivatives in good yields [14]. Presumably, these electrophiles react faster with the potassium derivative of isoprene than with KDA or KTMP occurring in the equilibrium mixture of the metallation. Other electrophiles, such as Me_3SiCl or Ch_3SSCH_3, give the isoprene derivatives in very low yields:

$$
\begin{array}{ccc}
CH_3 & & CH_2K \\
| & & | \\
H_2C=CCH=CH_2 + KDA & \rightleftarrows & H_2C=CCH=CH_2 + HDA \\
(KTMP) & & (HTMP) \\
k_1 \downarrow E & k_2 \downarrow E & \\
E\text{-}DA & H_2C=CCH=CH_2 & \\
(E\text{-}TMP) & | & \\
& CH_2E & k_2 \gg k_1 \text{ for} \\
& & R\text{-}Br \text{ or oxirane}
\end{array}
$$

When lithium bromide is added prior to the reactions with alkyl halides and oxirane, the deprotonation equilibrium shifts completely to the left side and functionalization of isoprene does not take place.

A similar strong influence of the counterion (Li^+ or K^+) on the metallation equilibrium is observed when α-methylstyrene, $PhC(CH_3)=CH_2$, is allowed to interact with LDA or with a 1:1 molar mixture of LDA and t-BuOK: whereas LDA does not cause any 'ionization', addition of the olefin to the mixture of

Table 4. Representative metallations of olefinic hydrocarbons[a]

Substrate (mol% excess)	Base–solvent system (mol% excess)	Temperatures (°C) (reaction time in min)
(cyclohexene with H, structure)	BuLi·t-BuOK, THF (no hexane used)	$-80(120) \rightarrow 0(0)$[b]
$CH_3CH=CH_2$ (300)	BuLi·t-BuOK, THF-hexane	$-80(15) \rightarrow 0(0)$
$(CH_3)_2C=CH_2$ (200)	BuLi·t-BuOK, THF-hexane	$-80(15) \rightarrow 0(0)$
$(CH_3)_2C=CH_2$	2 BuLi·t-BuOK, hexane (20)	$0(0) \rightarrow 20(180)$
$H_2C=C(CH_3)CH=CH_2$ (50)	LiTMP·t-BuOK, THF-hexane	$-70(10) \rightarrow -50(0)$
$Ph-C(CH_3)=CH_2$ (0)	LDA·t-BuOK, THF-hexane	$-70(10) \rightarrow -50(0)$
$CH_3CH=CHCH_3$ Z (300)	BuLi·t-BuOK, THF-hexane	$-80(15) \rightarrow 10(0)$
$H_2C=CHCH_2CH_3$ (300)	BuLi·t-BuOK, TMEDA-hexane	$-20(60) \rightarrow +10(0)$
$H_2C=CHCH_2C_4H_9$ (300)	BuLi·t-BuOK, THF-hexane	$-75(30) \rightarrow 0(0)$
(bicyclic structure with CH₃) (300)	BuLi·t-BuOK, TMEDA-hexane	$-20(60) \rightarrow +10(15)$
	BuLi·t-BuOK, hexane	$30(5) \rightarrow 50(60)$
(cyclohexene with CH₃ and isopropenyl substituent) (300)	BuLi·t-BuOK, THF-hexane	$-75(60) \rightarrow +20(0)$
(cyclohexadiene with H)	BuLi·TMEDA, hexane	$0(30) \rightarrow +5(30)$[c]
$PhCH_2CH=CH_2$ (0)	BuLi, THF-hexane (5)	$-10(10) \rightarrow 15(0)$
$PhCH_2CH=CH_2$	$NaNH_2$, liq. NH_3 (20)	$-33(0)$
$CH_3CH=CHCH=CH_2$ (20)	BuLi·t-BuOK, THF-hexane	$-80(5) \rightarrow -20(0)$
$H_2C=C(CH_3)C\equiv CH$ (dimetallation)	BuLi·t-BuOK, THF-hexane, 2 equiv. (10)	$-70(30) \rightarrow -5(0)$[d]
$CH_3(CH=CH)_2CH=CH_2$ (0)	KNH_2, liq. NH_3	$-33(0)$
Indene (0)	BuLi, THF-hexane	$-20(0)$
	BuLi, Et_2O-hexane	$-20(15)$
(0)	$LiNH_2$, liq. NH_3	$-33(0)$
Cyclopentadiene (0)	BuLi, THF or Et_2O-hexane	$<0(0)$
(5)	$LiNH_2$, liq. NH_3	$-40(0)$
(0)	Na, THF	$20-30(30)$
(0)	NaH, THF	$10-20(30)$

bases gives a deep-red solution, which reacts with oxirane to give the alcohol $PhC[(CH_2)_3OH]=CH_2$ in a good yield. The observed phenomena can be explained on the basis of the great difference between the ion-pair basicities of lithium- and potassium dialkylamides (see Sect. 2, Ref. [11]).

Compounds like cyclopentadiene, indene, and fluorene are considerably more acidic than the normal olefinic hydrocarbons. Their metallation can be conveniently achieved with a variety of basic reagents, including the 'conventional' ones, such as butyllithium and alkali amides in liquid ammonia [17, 18].

Literature

1. Hartmann J, Schlosser M (1976) Helv Chim Acta 59:453
2. Rauchschwalbe G, Schlosser M, (1975) Helv Chim Acta 58:1094
3. Andrianome M, Delmond B (1985) Tetrahedron Lett 6341
4. Schlosser M, Hartmann J, David V (1974) Helv Chim Acta 57:1567
5. Crawford RJ, Erman WF, Broaddus CD (1972) J Am Chem Soc 94:4298
6. Crawford RJ (1972) J Org Chem 37:3543
7. Akiyama S, Hooz J (1973) Tetrahedron Lett 4115
8. Bates RB, Beaves WA (1974) J Am Chem Soc 96:5001
9. Bahl JJ, Bates RB, Gordon III, B (1979) J Org Chem 44:2290
10. Seyferth D, Pornet J (1980) J Org Chem 45:1721
11. Brieger G, Anderson DW (1970) J Chem Soc, Chem Comm 1325
12. Gérard F, Miginiac P (1974) Bull Soc Chim France 1924
13. Bates RB, Gosselink DW, Kaczynski JA (1967) Tetrahedron Lett 199, 205
14. Klusener PAA, Hommes HH, Verkruijsse HD, Brandsma L (1985) J Chem Soc Chem Comm 1677
15. Schlosser M, Hartmann J (1973) Angew Chemie 85:544; (1973) Int ed (Engl) 12:508
16. Kloosterziel H, van Drunen JAA (1970) Recl Trav Chim Pays-Bas 89:368
17. Mallan JM, Bebb RL (1969) Chem Revs 69:693
18. Cedheim L, Eberson L (1973) Synthesis 159

5 Stereochemistry of Allylic Metallations

The allylic metal derivative formed by deprotonation of a 1-alkene or 2-alkene (at the most acidic allylic position) can occur in two conformations:

The ratio of the two forms can be determined by NMR spectroscopy [1] or—in many cases more easily—from the cis/trans ratio of the products $RCH=CHCH_2E$

[a] For experimental details see experimental section;
[b] → means: allowing the temperature of the reaction mixture to rise without cooling bath;
[c] At 20 °C or higher temperatures LiH is eliminated with formation of benzene;
[d] For experimental details see Brandsma L (1988) Preparative Acetylenic Chemistry, Elsevier.

formed by reaction with an electrophilic reagent, E, at the terminal carbon atom [2]. Reagents that attack preferentially at carbon atom 1, are $FB(OCH_3)_2$ and Me_3SiCl [3].

Form A and B can interconvert by rotation. The equilibrium ratio depends strongly on the nature of the group R (prim. alkyl, t-Bu, or Ph) and the counterion M^+ (Li^+, $MgBr^+$, Na^+, K^+, Cs^+). With primary alkyl groups and alkali-metal ions, form A is strongly favoured [1] but its isomer B with M = K can be specifically generated by treating the $trans$-2-alkene, alkyl-CH=CHCH$_3$ with the BuLi (or $sec.$ BuLi)·t-BuOK reagent at a sufficiently low temperature. A (M = K) is formed in a similar way from the cis-2-alkene: the interconversion between A and B is relatively slow when M = K. With pentadienylalkali-metal derivatives one can distinguish three forms [4]:

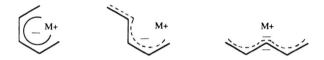

The preferred geometry of the anion, and hence the stereochemistry of the functionalization at the terminus, is determined by the counterion and by substituents present on the various carbon atoms.

Literature

1. Bates RB, Beavers WA (1974) J Am Chem Soc 96:5001; O'Brien DH, Russell CR, Hart AJ (1976) Tetrahedron Lett 37, and Refs mentioned in these papers
2. Schlosser M, Hartmann J, David V (1974) Helv Chim Acta 57:1567
3. Schlosser M, Fujita K (1982) Angew Chemie 94:320; (1982) Int ed (Engl) 21:309; (1982) Angew Chem Suppl 646 and previous papers of this group mentioned in this paper
4. Bosshardt H, Schlosser M (1980) Helv Chim Acta 63:2393

6 Dimetallation of Olefins

A variety of olefinic hydrocarbons have been dimetallated [1]. Most of the papers on this subject have been published by the groups of Klein and Bates. The usual base-solvent systems for the dimetallations are BuLi·TMEDA-hexane and BuLi·t-BuOK in hexane or homologues. In practice, clean dimetallations are not easily realizable. The result is very often a mixture of the mono- and dimetallated hydrocarbon and after reaction with some electrophilic reagent a corresponding mixture of mono- and disubstituted product is obtained. Examples of dimetallation reactions with preparative significance are the formations of the following species:

$$CH_2K$$
$$|$$
$$H_2C=C-C=CH_2$$ from 2,3-dimethylbutadiene and t-BuOK.BuLi in hexane [2]
$$|$$
$$CH_2K$$

K K   from isobutene and BuLi.t-BuOK in hexane [3]

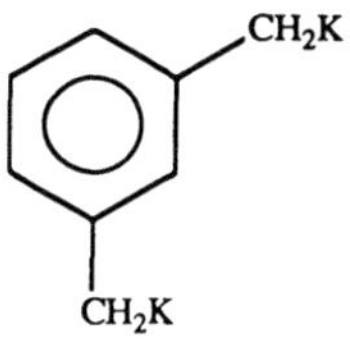

from m-xylene and BuLi.t-BuOK.TMEDA in hexane [4]

Literature

1. Bates RB (1980) 'Dianions and Polyanions' in: Buncel E, Durst T (eds) Comprehensive carbanion chemistry, part A, Elsevier
2. Bates RB, Gordon III B, Highsmith TK, White JJ (1984) J Org Chem 49:2981
3. Bates RB, Beavers WA, Gordon III B (1979) J Org Chem 44:3800; Klein J, Medlik A (1973) J Chem Soc Chem Comm 275
4. Unpublished results from the author's laboratory

7 Experiments

All reactions are carried out in an atmosphere of inert gas.
For general instructions concerning handling organoalkali reagents, see Vol. 1.
All temperatures are internal.

7.1 Metallation of Toluene with BuLi·t-BuOK in Hexane

$$PhCH_3 + BuLi \cdot t\text{-BuOK} \downarrow \xrightarrow[20 \to 40\,°C]{hexane} PhCH_2K \downarrow + t\text{-BuOLi}$$

$$PhCH_2K + Me_3SiCl \xrightarrow[-40 \to +15\,°C]{hexane, Et_2O} PhCH_2SiMe_3 + KCl$$

Apparatus: Reaction vessel as shown in Fig. 1 below; 1 l. (for a detailed description see Vol. 1, p. 8)

Scale: 0.5 molar

Introduction

Lochmann [1] and Schlosser [2] obtained benzylpotassium by treating toluene with equimolar amounts of BuLi and t-BuOK in hexane. Schlosser removed the

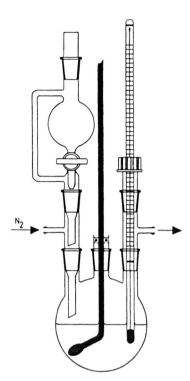

soluble lithium *tert*-butoxide by suction filtration and washing of the red solid with toluene. In many derivatization reactions the presence of *t*-BuOLi can be tolerated, however. In our procedure for the metallation of toluene and subsequent trimethylsilylation *t*-BuOLi is not removed, while the excess of toluene is limited to 300% (instead of ∼ 800% used by Schlosser) by using more hexane.

Procedure

Dry toluene (200 ml), hexane (50 ml), and *t*-BuOK (0.50 mol, 56.5 g, powdered under a blanket of inert gas) are placed in the flask. A solution of 0.50 mol of BuLi in ∼ 330 ml of hexane is added over 15 min. After this addition, which occurs without external cooling, the red suspension is stirred for an additional 45 min at 40 °C. Et_2O (200 ml) is then added while the suspension is cooled in a bath with dry ice and acetone. Freshly distilled Me_3SiCl (0.65 mol, 70.6 g) is then added over 15 min while keeping the temperature between − 40 and − 20 °C. After the addition, the cooling bath is removed and stirring is continued for a few min until the colour has disappeared. The reaction mixture is poured into a cold (0 °C) mixture of 50 ml 36% hydrochloric acid and 400 ml of water. After vigorous shaking, the upper layer is washed four times with dilute hydrochloric acid in order to remove *t*-butylalcohol. The organic solution is dried over $MgSO_4$, after which the greater part of the Et_2O and hexane is distilled off at atmospheric pressure through a 30-cm Widmer column. After cooling the distillation flask to below 30 °C, a very careful vacuum distillation is carried out. Benzyl trimethylsilane, b.p. 68 °C/12 mmHg, $n_D(20)$ 1.4924, is obtained in $\sim 90\%$ yield.

Literature

1. Lochmann L, Pospísil J, Lím D (1966) Tetrahedron Lett 257
2. Schlosser M, Hartmann J (1973) Angew Chemie 85:544; (1973) Int ed (Engl) 12:508

7.2 Metallation of 1- and 2-Methylnaphthalene with BuLi·*t*-BuOK·TMEDA in Hexane

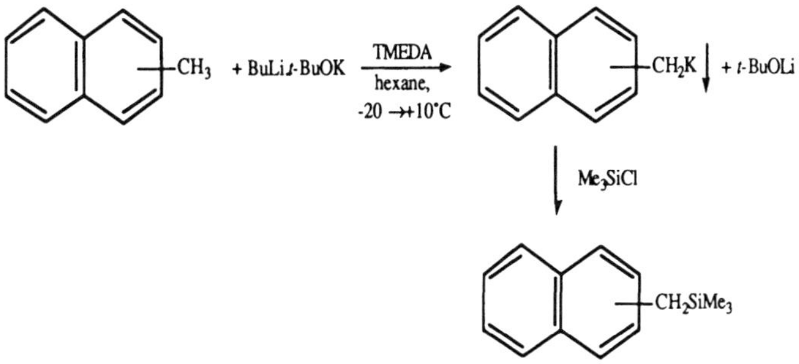

Apparatus: p. 24, 500 ml

Scale: 0.10 molar

Introduction

Schlosser and Hartmann (see exp. 1) α-metallated 2-methylnaphthalene with an equivalent amount of BuLi·*t*-BuOK in hexane and obtained the carboxylic acid in a good yield by reaction of the lithium compound (formed by addition of anhydrous lithium bromide) with carbon dioxide. Using the procedure of Schlosser and Hartmann we were unable to bring about complete conversion of the methylnaphthalenes. Subsequent reaction with trimethylchlorosilane gave a mixture of the starting compounds and the silyl derivative (ratios varying from about 1:3 to 1:5). Separation by distillation is very difficult, since the difference in boiling points is relatively small. Similar problems may arise after functionalizations with other types of electrophilic reagents.

The advantage of our variant of the Lochmann-Schlosser reagent [1] is that the α-metallations can proceed under almost homogeneous conditions: the BuLi·*t*-BuOK·TMEDA reagent has a reasonable solubility in hexane, under these conditions a complete conversion of the methylnaphthalenes can be attained with an excess of the basic reagent. Using the BuLi·TMEDA reagent we obtained mixtures of the (trimethylsilylmethyl)naphthalenes and ring-silylated isomers. The potassiomethylnaphthalenes obtained by our procedure can be easily converted into the lithio compounds by adding a solution of lithium bromide in THF.

Procedure

Butyllithium (0.06 mol in 37 ml of hexane), TMEDA (0.07 mol, 8.1 g) and an additional volume of 30 ml of hexane are placed in the flask. t-BuOK (0.06 mol, 6.7 g, powdered under inert gas) is added with stirring at a moderate rate and cooling between − 30 and − 40 °C. After an additional to min (at − 30 °C) 0.05 mol (7.1 g) of 1- or 2-methylnaphthalene dissolved in 25 ml of hexane is added in one portion. The dark-red suspension is stirred for 45 min at − 20 °C, then the cooling bath is removed and the temperature allowed to rise to + 10 °C. After an additional 10 min THF (50 ml) is added (note 1) and stirring at 10 °C is continued for 15 min, then the suspension is cooled to − 30 °C and freshly distilled trimethylchlorosilane (0.10 mol, 10.9 g) is added with vigorous stirring. When the red colour has disappeared completely, 100 ml of 2N hydrochloric acid is added with vigorous stirring. The organic layer and one ethereal extract of the aqueous layer are dried over MgSO$_4$ and subsequently concentrated under reduced pressure. Distillation of the remaining liquid through a 10-cm Vigreux column gives, after a small first fraction (partly starting compound), the α-silylated methylnaphthalenes, b.p. ∼ 150 °C/18 mm Hg, in ∼ 85% yields.

Notes

1. The excess of base that has not yet reacted with TMEDA [2], is destroyed by reaction with THF.

Literature

1. Brandsma L, Verkruijsse HD, Schade C, von R Schleyer P (1986) J Chem Soc Chem Comm 260
2. Köhler FR, Hertkorn N, Blümel J (1987) Chem Ber 120:2081

7.3 α-Metallation of Ethylbenzene

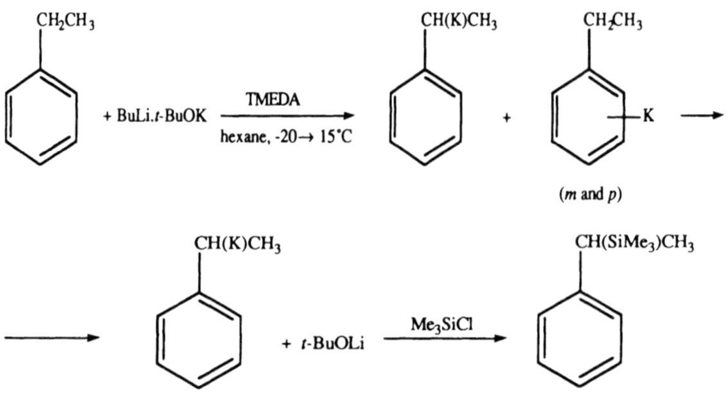

Apparatus: p. 24, 500 ml

Scale: 0.05 molar

Introduction

Benkeser et al. [1] treated ethylbenzene with n-pentylsodium and obtained after 20 hours a mixture of PhCH(Na)CH$_3$ and (smaller amounts of) m- and p-ethylphenylsodium. Shorter interaction times gave predominantly the ring-metallated products. Crimmins and Rather [2] reinvestigated the reaction: using a mixture of n-pentylsodium and TMEDA they found a complete conversion of ethylbenzene into the α-metallated product within 1 h. If TMEDA was omitted, the m- and p-metallated products were found, in addition to PhCH(Na)CH$_3$.

The preparation of n-pentylsodium (or -potassium) either from n-pentyl chloride and a metal dispersion or from dipentylmercury and the alkali metals is a laborious procedure. Therefore we investigated the metallation of ethylbenzene with a 1:1:1 molar mixture of t-pentyl-ONa (commercially available), BuLi and TMEDA in hexane [3]. The various sodium compounds in the reaction mixture were trapped by quenching with trimethylchlorosilane. Our results and those of Benkeser c.s. show a strong resemblance. Even after 2 h at 15 °C and with a 300 mol% excess of ethylbenzene the mixture still contained considerable amounts of ring-metallated intermediates. Comparison of our results with those reported by Crimmins and Rather strongly suggests that the mixture of n-pentylsodium and TMEDA is effectively a stronger base than t-pentyl-ONa · BuLi · TMEDA (compare Ref. [4]).

In the procedure described below, the system t-BuOK · BuLi · TMEDA is used. Although comparison with the reaction conditions of Crimmins [2] shows that our basic system is less reactive, our procedure is more convenient since it uses commercially available reagents.

Procedure

Ethylbenzene (0.2 mol, 21.2 g), TMEDA (0.07 mol, 8.1 g), hexane (50 ml), and t-BuOK (0.07 mol, 7.9 g, powdered under inert gas) are placed in the flask. BuLi (0.055 mol in 36 ml of hexane, 0.005 mol excess to compensate for traces of impurities and moisture present in the reagents and in the reaction flask) is added over a few min from a syringe while stirring the mixture at ~ − 30 °C. After 45 min at − 20 °C, the cooling bath is removed and stirring at + 15 °C is continued for 2.5 h, then 100 ml of Et$_2$O is added. The mixture is cooled to − 30 °C and freshly distilled trimethylchlorosilane 0.07 mol, 7.6 g) is added in one portion with vigorous stirring. Water (100 ml) is added after the colour has disappeared, after which the layers are separated and the aqueous layer extracted once with Et$_2$O. After drying the solution over MgSO$_4$, the solvent is removed in vacuo and the remaining liquid carefully distilled through a 30-cm Vigreux column. The α-silylated product, b.p. 82 °C/12 mmHg, n$_D$(20) 1.4982, is obtained in greater than 80% yield.

Literature

1. Benkeser RA, Trevillyan AE, Hooz J (1962) J Am Chem Soc 84:4971
2. Crimmins TF, Rather EM (1978) J Org Chem 43:2170
3. Verkruijsse HD, Brandsma L (unpublished results)
4. Lehmann R, Schlosser M (1984) Tetrahedron Lett 745

Note

Toluene, *o*-, *m*-, and *p*-xylene, mesitylene, and higher homologues of ethylbenzene may be α-metallated by a similar procedure.

7.4 α,α′-Dimetallation of *m*-Xylene

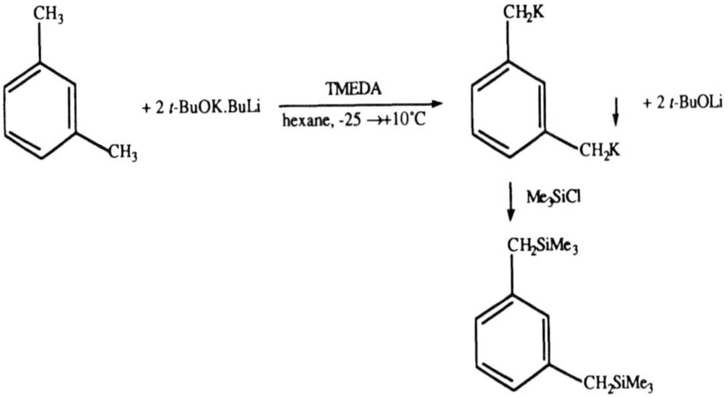

Apparatus: p. 24, 500 ml

Scale: 0.05 molar (*m*-xylene)

Introduction

The successful dimetallations of *o*- and *m*-xylene and of some dimethylnaphthalenes with a mixture of butylsodium and TMEDA reported by Trimitsis et al. [1] led us to investigate the possibility of using our more readily available reagents BuLi·*t*-BuONa·TMEDA and BuLi·*t*-BuOK·TMEDA (compare Vol. 1, p. 19). With the first mentioned combination only mono-(α-)metallation could be achieved after a reaction time of about 2 h at −20 °C, but a reasonably satisfactory result was obtained with the *t*-BuOK-containing reagent. Using two mol equivalents plus 30 mol% excess, a 80:20 mixture of α,α′-disilylated and α-(mono)silylated products was obtained after quenching with trimethylchlorosilane. With a large excess of base this ratio will undoubtedly become somewhat higher, but the excess will mainly react with the TMEDA to cause 1,2-elimination of Me_2NH and subsequent α-metallation of $H_2C=CHNMe_2$ [2]. For certain reactions with electrophiles it may be desired to destroy the excess of basic reagent and to exchange potassium in the generated intermediate for lithium. This can be done by successively adding at 0 °C a sufficient amount of THF and a solution of anhydrous lithium bromide in THF. The THF reacts with the excess of the base to give ethene and $H_2C=CHOK$ by a cyclo-elimination. Being a weakly basic nucleophile, the enolate of acetaldehyde is not likely to interfere during subsequent reactions with most electrophiles.

Thus, α,α'-dimetallation of m-xylene (and presumably also of o-xylene, but not of p-xylene (compare Ref. [3]) can be achieved in a convenient way using commercially available reagents.

We have also carried out the procedures described by Bates and co-workers [3]: a mixture of m-xylene, 2.2 equivalents of finely powdered t-BuOK, 2.2 equivalents of BuLi and hexane was heated under reflux for 2.5 h, then THF and excess of Me_3SiCl were successively added with cooling below 0 °C. After a very sluggish reaction (heating to 30–35 °C was necessary) the expected α,α'-disilylated derivative was obtained in ~ 60% yield. Shorter periods of reflux (1 h, as mentioned by Bates) gave mainly the mono-silyl derivative.

Procedure

A mixture of finely powdered (under inert gas) t-BuOK (0.15 mol, 17.0 g), TMEDA (0.15 mol, 17.4 g), and hexane (50 ml) is placed in the flask. A solution of 0.13 mol of butyllithium in 86 ml of hexane is added over a few min from a syringe while cooling the mixture below − 30 °C. After stirring (at a moderate rate) for 15 min at − 30°C, a light-yellow solution has formed while the greater part of the t-BuOK has passed in solution. m-Xylene (0.05 mol, 5.3 g) is added in one portion. The temperature rises to above − 25 °C upon this addition, but is kept close to − 25 °C by cooling. The yellow-orange suspension which is formed immediately, is stirred for 1.5 h at − 20 °C, then the cooling bath is removed and the temperature allowed to rise to 10–15 °C. Stirring is continued for half an hour, then THF (80 ml) is added. The suspension is kept for 10 min at ~ + 10 °C (during this period the excess of BuLi.t-BuOK reacts with the THF), then it is cooled to − 40 °C and 0.15 mol (16.3 g) of freshly distilled trimethylchlorosilane is added in one portion with vigorous stirring. The cooling bath is removed and, after the colour has disappeared, 100 ml of ice water is added. The aqueous layer is extracted twice with pentane, after which the combined organic solutions are washed with cold dilute (2 to 3 N) hydrochloric acid in order to remove TMEDA and t-BuOH. After drying the solution over $MgSO_4$, the solvent is removed under reduced pressure and the remaining liquid carefully distilled through a 20-cm Vigreux column. After a small first fraction (mainly the mono-silyl derivative), the expected disilyl derivative, b.p. ~ 125 °C/12 mmHg, $n_D(20)$ 1.4904, is obtained in ~ 80% yield.

Literature

1. Trimitsis GB, Tuncay A, Beyer RD, Ketterman KJ (1973) J Org Chem 38:1491
2. Köhler FH, Hertkorn N, Blümel J (1987) Chem Ber 120:2081
3. Bates RB, Ogle CA (1982) J Org Chem 47:3949

7.5 Lithiation of Toluene, Xylene, and Mesitylene with BuLi·TMEDA

CH₃ [structure] —CH₃ + BuLi.TMEDA $\xrightarrow[30°C \to reflux]{hexane}$ CH₂Li.TMEDA [structure] —CH₃ + C₄H₁₀ ↑

(Analogous for mesitylene)

Subsequent reactions with dimethylformamide, dimethyldisulfide, methyl isothiocyanate, carbon dioxide, trimethylchlorosilane, paraformaldehyde, cyclohexyl bromide, and bromoacetaldehyde diethylacetal.

Apparatus: For the metallation a 500-ml round-bottomed, three-necked flask is used, equipped with a gas-inlet, a mechanical stirrer, and a reflux condenser, connected with a washing bottle filled with parrafin oil (the inner tube should not dip more than 0.5 cm into the oil). All connections are made gas-tight. For the functionalizations the apparatus is modified as indicated below.

Scale: 0.10 molar

Introduction

Metallation of toluene, the isomeric xylenes, and mesitylene can be readily achieved with BuLi·t-BuOK or with BuLi·t-BuOK · TMEDA in hexane [1]. Some derivatization reactions, particularly those with carbonyl compounds and carbon dioxide give much better results with the lithium compounds. In these cases the potassiated aromatic compounds have to be treated first with anhydrous lithium bromide in order to exchange potassium for lithium. The α-lithiated aromatic compounds can also be obtained directly by treating the substrates with BuLi · TMEDA in hexane at elevated temperatures [3]. The use of a large excess of the aromatic compound allows the metallation to be completed within 15 to 30 min, furthermore attack of TMEDA by BuLi is reduced to a minimum. With toluene this procedure gives a mixture of ∼ 90% α-lithiated and ∼ 10% ring-lithiated products, but the α-lithiation of the xylenes and mesitylene proceeds with a selectivity of at least 96%. 1- and 2-methylnaphthalene undergo α- and ring-metallation to a comparable extent.

Procedure

Toluene (0.3 mol, 28 g), o-, m-, or p-xylene (0.3 mol, 32 g) or mesitylene (0.2 mol, 24 g) and TMEDA (0.11 mol, 12.8 g) are placed in the flask. A solution of 0.10 mol of butyllithium in 66 ml of hexane is added from a syringe. During this operation the temperature rises to ca. 30 °C. The solutions are heated under gentle reflux until the evolution of gas has stopped completely. This takes 20 to 30 min. The red solutions are then cooled to room temperature and THF (70 ml) is added. The reflux condenser is replaced with a thermometer-gas outlet combination (see p. 24). The functionalization reactions are carried out as follows.

Paraformaldehyde

Dry powdered paraformaldehyde (4.5 g, excess; dried by heating the powder for 30 min at 60 to 70 °C in a vacuum of 10 to 20 mmHg) is added in one portion at 15 °C to the solution of mesityllithium. The temperature rises gradually but is kept close to 30 °C by occasional cooling. After about 1.5 h (at 30 °C, the orange colour has practically disappeared) the reaction mixture is cooled to 10 °C and 150 ml of 3 N hydrochloric acid is added with vigorous stirring. The layers are separated and the aqueous layer is extracted twice with Et_2O. After being washed with water, the organic solution is dried over $MgSO_4$ and subsequently concentrated under reduced pressure. Distillation of the remaining liquid through a 15-cm Vigreux column gives (after a volatile fraction of mesitylene) the alcohol, b.p. $\sim 100\,°C/0.5\,mmHg$, in $\sim 70\%$ yield.

Dimethylformamide

After addition of 100 ml of Et_2O, the solution of p-xylyllithium is cooled to $-90\,°C$ (bath with liquid N_2, occasional cooling). Dimethylformamide (10 g, excess) is then added in one portion with vigorous stirring and the cooling bath is removed. After a few seconds the orange colour has disappeared. The reaction mixture is poured (with continuous swirling by hand) into a cold (0 °C) mixture of 40 g of 30% hydrochloric acid and 300 ml of water. After vigorous shaking in a separatory funnel, the layers are separated. The aqueous layer is extracted twice with small portions of pentane. The combined organic solutions are washed three times with a concentrated aqueous solution of ammonium chloride and subsequently dried over $MgSO_4$. After removal of the solvent in vacuo the remaining liquid is distilled through a 20-cm Vigreux column to give the aldehyde, b.p. 100 °C/15 mmHg, $n_D(20)$ 1.523 in greater than 80% yield. The aldehyde polymerizes (trimer?) easily: depolymerization can be effected by distillation (strong heating is necessary) from a few grams of potassium hydrogen sulfate.

Dimethyldisulfide

After addition of 50 ml of Et_2O, the solution of o-xylyllithium is cooled to $-60\,°C$ and dimethyldisulfide (0.12 mol, 11.3 g) is added in one portion with vigorous stirring. The cooling bath is removed and, after the orange colour has disappeared, 100 ml of water is added with vigorous stirring. The aqueous layer is extracted twice with small portions of pentane or Et_2O, the combined organic solutions are washed three times with small portions of water and subsequently dried over $MgSO_4$. The product (b.p. 107 °C/18 mmHg, yield > 80%) is isolated in the same way as the aldehyde.

Methyl isothiocyanate

The solution of *p*-xylyllithium is cooled to − 80 °C (bath with liquid N_2, occasional cooling) after which a mixture of methyl isothiocyanate (0.11 mol, 8.0 g) and 30 ml of THF is added in one portion with vigorous stirring. The cooling bath is removed and the temperature allowed to rise to − 10 °C. 3 N Hydrochloric acid (120 ml) is then added with vigorous stirring. The layers are separated as completely as possible: addition of a sufficient amount of chloroform may be necessary to keep the thioamide in solution. After drying the organic solution over $MgSO_4$, the solvent is removed under reduced pressure. The remaining solid (yield ∼ 90%), reasonably pure thioamide $CH_3—C_6H_4—CH_2C(=S)NHCH_3$, is crystallized from Et_2O to give (after one or two recrystallizations) the pure thioamide, m.p. 87.5 °C.

Slow addition of methyl isothiocyanate might lead to metallation of the initial product $ArCH_2—C(SLi)=NCH_3$, resulting in the formation of $ArCH(Li)C(SLi)=NCH_3$.

Trimethylchlorosilane

The solution of *m*-xylyllithium is cooled to ∼ − 40 °C and freshly distilled trimethylchlorosilane (0.12 mol, 13.0 g) is added in one portion with vigorous stirring. After the work-up, carried out as described for the reaction with CH_3SSCH_3, the product (b.p. 85 °C/10 mmHg, $n_D(20)$ 1.4876) is isolated in an excellent yield by careful distillation.

Cyclohexyl bromide

The solution of benzyllithium (containing ∼ 10% of ring-lithiated intermediates, which give no coupling product with cyclohexyl bromide) is cooled to − 40 °C and cyclohexyl bromide (0.08 mol, 13.0 g) is added over a few min. The cooling bath is then removed and the temperature allowed to rise to 0 °C. After the usual work-up the product (b.p. 117 °C/12 mmHg, $n_D(20)$ 1.5179) is isolated in ∼ 60% yield by careful distillation.

Carbon dioxide

The procedure for the carboxylation of *p*-xylyllithium is carried out in a way similar to the carboxylation of $(CH_3S)_2CHLi$, described in Chap. III, Exp. 8. The solution of *p*-xylyllithium in THF and hexane is precooled to ca. − 20 °C. The carboxylic acid is obtained in ∼ 80% yield. For the isolation and purification (recrystallization from a 1:2 Et_2O-hexane mixture) see Chap. III Exp. 8. The m.p. is 89.7 °C.

Bromoacetaldehyde diethylacetal

Reaction of $BrCH_2CH(OC_2H_5)_2$ (0.10 mol, 19.7 g) with the solution of $PhCH_2Li$ in THF and hexane under the conditions described for cyclohexyl bromide gives $PhCH_2CH_2CH(OC_2H_5)_2$, b.p. 135 °C/12 mmHg, $n_D(20)$ 1.4961 (yield ∼ 50%), after a large, more volatile fraction of the bromoacetal.

Literature

1. Schlosser M, Hartmann J (1973) Angew Chemie 85:544; (1973) Int ed (Engl) 12:508; Lochmann L, Pospisil J, Lim D (1966) Tetrahedron Lett 257; Brandsma L, Verkruijsse HD, Schade C, von R Schleyer P (1986) J Chem Soc Chem Comm 260
2. Köhler FH, Hertkorn N, Blümel J (1987) Chem Ber 120:2081
3. Eberhardt GG, Butte WA (1964) J Org Chem 29:2928

7.6 Metallation of Propene and Isobutene

$(CH_3)_2C{=}CH_2 + BuLi \cdot t\text{-}BuOK$
(analogous for propene)

$\xrightarrow[-80\rightarrow 0\,°C]{\text{THF-hexane}} KCH_2{-}C(CH_3){=}CH_2 + t\text{-}BuOLi + C_4H_{10}$

$\xrightarrow[-25\rightarrow +10\,°C]{\text{TMEDA-hexane}} KCH_2{-}C(CH_3){=}CH_2 + t\text{-}BuOLi + C_4H_{10}$

$KCH_2{-}C(CH_3){=}CH_2 + LiBr$
(analogous for allylpotassium)

$\xrightarrow[-10\rightarrow 0\,°C]{\text{THF-hexane}} LiCH_2{-}C(CH_3){=}CH_2 + KBr\downarrow$

Subsequent reactions with n-octyl bromide, c-hexyl bromide, bromoacetaldehyde diethylacetal, benzyl chloride, oxirane, carbon dioxide, carbon disulfide (followed by reaction with methyl iodide), and methyl isothiocyanate.

Apparatus: p. 24, 500 ml. The carboxylation reaction is carried out in a 1-1 round-bottomed, four-necked flask, equipped with a gas-inlet tube, dipping in the solution, a mechanical stirrer, a thermometer and a gas-outlet. The latter is removed when the solution of isobutenyllithium is introduced by means of a syringe (internal diameter of the needle ~ 1 mm).

Scale: 0.10 molar

Introduction

For the metallation of propene and homologues the following base-solvent systems are in principle available: BuLi·TMEDA-hexane [1], BuLi·t-BuOK-hexane [2], BuLi·t-BuOK-THF + hexane [3], BuLi·t-BuOK·TMEDA-hexane [4]. In order to maintain a sufficiently high concentration, the metallation of gaseous alkenes (propene, 1- and 2-butene, and isobutene) has to be carried out at relatively low temperatures. As a consequence, reaction times may be long [1]. With the third and fourth base-solvent system the gaseous alkenes can be metallated quickly, while reactions with a number of electrophiles indicate that the metallations have proceeded successfully.

Procedure

1. Metallation with BuLi · t-BuOK in THF-hexane

A solution of 0.10 mol of BuLi in 66 ml of hexane is placed in the flask. After cooling to $-80\,°C$, liquified propene (0.4 mol, ~ 16.8 g) or isobutene (0.3 mol, ~ 17 g) is added in one portion. A solution of 0.10 mol (11.3 g) of t-BuOK in 80 ml of THF is added dropwise over 15 min, while the temperature is kept between -70 and $-80\,°C$. After an additional 15 min, the cooling bath is removed and the temperature allowed to rise to $0\,°C$. To prepare the lithium compounds, a solution of 0.10 mol (8.7 g) of anhydrous lithium bromide (the commercial anhydrous salt is heated for 30 min at $150\,°C$ in a vacuum of 15 mmHg or less) in 40 ml of THF is added over a few seconds. A fine white suspension is formed.

2. Metallation with BuLi · t-BuOK · TMEDA in hexane

A solution of 0.10 mol of BuLi in 66 ml of hexane is placed in the flask which is completely immersed in the cooling bath. Finely powdered t-BuOK (0.10 mol, 11.3 g) is added over a few s with cooling at $\sim -30\,°C$. TMEDA (0.12 mol, 14.0 g) is added over a few min at the same temperature. The mixture is stirred for about 15 min at $-30\,°C$ during which period most of the solid passes into solution. Isobutene, 1-butene, or Z-2-butene (0.4 mol, 22.4 g, liquified in a trap) is added in one portion at -25 to $-30\,°C$. The yellow to orange suspension is stirred for 1 h at $-20\,°C$, then the cooling bath is removed and the temperature allowed to rise to $+10\,°C$. THF (70 ml) is then added.

3. Alkylation with n-$C_8H_{17}Br$, $PhCH_2Cl$, c-$C_6H_{11}Br$, $BrCH_2CH(OC_2H_5)_2$

a. *n-Octyl bromide* (0.08 mol, 15.4 g) is added over a few s at $-60\,°C$ to the solution of isobutenylpotassium in THF and hexane, after which the cooling bath is removed. The temperature rises over a few s to $10\,°C$ or higher. After an additional 15 min, the reaction mixture is poured into cold 3 to 4 N hydrochloric acid (a sufficient amount should be used to bring the pH of the aqueous layer at 6 or somewwhat lower). The aqueous layer is extracted with Et_2O after which the organic solutions are washed with water and dried over $MgSO_4$. The solvent is then removed under reduced pressure and the remaining liquid distilled through a 20-cm Vigreux column to give $H_2C{=}C(CH_3)C_9H_{19}$, b.p. $85\,°C/14$ mmHg, $n_D(20)$ 1.4335, in $\sim 80\%$ yield.

b. *Benzyl chloride* (0.08 mol, 10.1 g) is added dropwise over a few min at $\sim -70\,°C$, then the cooling bath is removed. When the temperature of the mixture has reached $0\,°C$, dilute hydrochloric acid is added and the product isolated as described above. $PhCH_2CH_2C(CH_3){=}CH_2$, b.p. $75\,°C/12$ mmHg, $n_D(20)$ 1.5069, is obtained in $\sim 85\%$ yield.

c. *Cyclohexyl bromide* (0.08 mol, 13.0 g) is added at $\sim -50\,°C$ after which the temperature is allowed to rise to $+10\,°C$. $H_2C{=}C(CH_3)CH_2$-c-C_6H_{11}, b.p. $55\,°C/14$ mmHg, $n_D(20)$ 1.4545, is obtained in $\sim 70\%$ yield.

d. *Bromoacetaldehyde diethylacetal* (0.08 mol, 15.8 g) is added at $\sim -50\,°C$, after which the temperature is allowed to rise to $+10\,°C$. $H_2C= C(CH_3)CH_2CH_2CH(OC_2H_5)_2$, b.p. $65\,°C/12\,mmHg$, is obtained in $\sim 80\%$ yield.

e. *Hydroxyalkylation with oxirane.* A mixture of 0.14 mol (6.2 g) of oxirane and 40 ml of THF is added dropwise at $-40\,°C$ over 15 min to the solution of isobutenylpotassium, then the cooling bath is removed and the temperature allowed to rise to $-10\,°C$ (note). The reaction mixture is hydrolysed with cold 3 to 4 N hydrochloric acid (a sufficient amount to bring the pH of the aqueous phase at ~ 5). The aqueous layer is extracted 6 times with small portions of Et_2O. The (unwashed) combined organic solutions are dried over $MgSO_4$, after which the greater part of the solvent is distilled off at atmospheric pressure through a 40-cm Vigreux column (bath temperature not higher than $110\,°C$). Careful distillation of the remaining liquid gives $H_2C=C(CH_3)(CH_2)_3OH$, b.p. $40\,°C/12\,mmHg$, $n_D(20)$ 1.4388, in $\sim 70\%$ yield.

Note: The temperature should not be allowed to rise above $-10\,°C$, since the excess of oxirane may react with t-BuOLi to afford t-BuOCH$_2$CH$_2$OH which is not easy to separate from the desired product.

f. *Carboxylation.* THF (100 ml) is placed in the flask. After cooling to below $-80\,°C$, a rapid stream of carbon dioxide is introduced, while keeping the temperature between -80 and $-90\,°C$. The suspension of $H_2C= C(CH_3)CH_2Li$ and KBr (method 1) is added over ~ 7 min by means of a syringe, while continuing the rapid flow of CO_2 (for the procedure see Chap. III-3, Exp. 8). The product $H_2C=C(CH_3)CH_2COOH$, b.p. $\sim 80\,°C/20\,mmHg$, $n_D(20)$ 1.430, is obtained in $\sim 65\%$ yield.

g. *Reaction with carbon disulfide* (compare Ref. [3]). The suspension of allyllithium or isobutenyllithium and potassium bromide (see first method) is cooled to $-110\,°C$, after which carbon disulfide (0.10 mol, 7.6 g) is added in one portion with vigorous stirring. The temperature of the brown solution rises to about $-70\,°C$. The cooling bath is then removed and the temperature allowed to rise to $-20\,°C$. The reaction mixture is cooled again to $-65\,°C$ and methyl iodide (0.13 mol, 18.5 g) is added in one portion. The cooling bath is removed and the temperature allowed to rise to $10\,°C$. Water (100 ml) is then added with vigorous stirring. The aqueous layer is extracted twice with Et_2O or pentane. The combined organic solutions are washed with water, dried over $MgSO_4$, and subsequently concentrated under reduced pressure. Careful distillation of the remaining liquid through a 20-cm Vigreux column gives the ketene-S,S-acetals $H_2C=CHCH=C(SCH_3)_2$, b.p. $\sim 50\,°C/1\,mmHg$, $n_D(20)$ 1.6038 and $H_2C= C(CH_3)CH=C(SCH_3)_2$, b.p. $\sim 50\,°C/0.5\,mmHg$, $n_D(20)$ 1.5861, in greater than 75% yields.

h. *Reaction with methyl isothiocyanate.* A mixture of methyl isothiocyanate (0.10 mol, 7.3 g) and 20 ml of THF is added over a few s to the suspension of allyllithium or isobutenyllithium and potassium bromide (method 1), which has been pre-cooled to $-90\,°C$. The resulting reaction mixture (yellowish suspension) is stirred for an additional 10 min at $-60\,°C$, then a mixture of 20 g of

36% hydrochloric acid and 150 ml of water is added with vigorous stirring and without external cooling. The aqueous layer is extracted once with Et_2O, after which the combined organic solutions are washed twice with water and subsequently dried over $MgSO_4$. The solvent is removed under reduced pressure and the remaining product purified by distillation through a short column. $H_2C=CHCH_2C(=S)NHCH_3$, b.p. ~ 95 °C/1.5 mmHg, $n_D(20)$ 1.5698 and $H_2C=C(CH_3)CH_2C(=S)NHCH_3$, b.p. ~ 110 °C/1.5 mmHg, $n_D(20)$ 1.5595 are obtained in ~ 80% yields.

Literature

1. Akiyama S, Hooz J (1973) Tetrahedron Lett 4115
2. Schlosser M, Hartmann J (1973) Angew Chemie 85:544; (1973) Int ed (Engl) 12:508
3. Heus-Kloos YA, de Jong RLP, Verkruijsse HD, Brandsma L, Julia S (1985) Synthesis 958
4. Brandsma L, Verkruijsse HD, Schade C, von R Schleyer P (1986) J Chem Soc Chem Comm 260

7.7 Metallation of Various Olefins with Strongly Basic Reagents (Compare Table 4)

Apparatus: p. 24, 500 ml

Scale: 0.10 molar

Introduction

In Table 4 reaction conditions for the metallation of a number of olefinic compounds, as derived from our investigations, are summarized. The deprotonation of non-conjugated olefins proceeds rather sluggishly and satisfactory rates of conversion can be attained only when the substrate is used in a large excess. Under these conditions attack of the solvent (THF!) by the base is suppressed.

In most cases the excess of starting compound can be readily separated from the derivatization product by distillation. Problems may arise, however, when a substrate with a relatively high molecular weight is to be metallated and subsequently functionalized, since the difference in the boiling points of the starting compound and end product is then much smaller. In such a case one may consider to perform the metallation with an excess of *base* (if there is no chance on dimetallation or other complications) and to destroy the excess by reaction with THF at somewhat higher temperatures (resulting in the formation of $H_2C=CH_2$ and $H_2C=CH-OM$) prior to the further synthetic operations. Such a consideration also makes sense when the substrate is expensive or not readily available.

In this experiment a number of metallations in Table 4 are described in somewhat more detail.

Procedure

1-Heptene, 1-butene, Z-2-butene

a. A solution of 0.105 mol of BuLi in 66 ml of hexane is placed in the flask. Freshly distilled 1-heptene (0.4 mol, $\sim$ 40 g), liquified 1-butene, or Z-2-butene (0.4 mol, 22.5 g cooled below $-30°$ and then diluted with 25 ml of cold hexane) is added in one portion at $-85°C$ (occasional cooling with liquified N_2), after which a solution of 0.10 mol (11.3 g) of t-BuOK in 80 ml of THF is added dropwise over 15 min to the gently stirred mixture. During this addition the temperature of the yellow to orange mixture is maintained between -80 and $-85°C$. After an additional period (see Table 4), the cooling bath is removed and the temperature allowed to rise to the level indicated in the Table. Reactions with electrophilic reagents (Me_3SiCl, CH_3SSCH_3, alkyl bromides, or oxirane are added over a few min after cooling to $-50°C$) are carried out without delay and generally give (with high yields) mixtures of products derived from both mesomeric forms of the 'anion' of the alkene. Addition (at $-40°C$) of a 5% excess of Me_3SiCl to the solution of metallated heptene, followed by hydrolysis, extraction of the organic layer with Et_2O, washing of the organic solution with 3 N hydrochloric acid, and careful distillation gives a mixture of silyl derivatives, b.p. $63-67°C/$ 15 mmHg, in $\sim$ 75% yield. The components are Z-C_4H_9CH=CH—CH_2SiMe_3 ($\sim$ 88%) and (comparable amounts of) E-C_4H_9CH=$CHCH_2SiMe_3$, and $C_4H_9CH(SiMe_3)CH$=CH_2.

b. A solution of 0.105 mol of BuLi in 69 ml of hexane is placed in the flask. Finely powdered t-BuOK (0.10 mol, 11.3 g, powdered in an inert atmosphere) is introduced over a few s while keeping the temperature between 0 and $-10°C$. After gentle stirring for 15 min, the suspension is cooled to $-40°C$ and TMEDA (0.12 mol, 13.9 g) is added in one portion. The mixture is gently stirred at $-30°C$ for $\sim$ 15 min, after which period most of the suspended material has dissolved. Subsequently, Z-2-butene (0.4 mol diluted with 30 ml of strongly cooled hexane) or 1-heptene is added over a few s, while keeping the temperature between -20 and $-30°C$. The orange or red mixture is stirred at $-20°C$ (the greater part of the flask being immersed in the cooling bath) for the period indicated in Table 4, then the cooling bath is removed and the temperature allowed to rise to $+10°C$. THF (80 ml) is then added with cooling below $0°C$, after which reactions with electrophilic reagents are carried out.

Limonene

Freshly distilled limonene (300% excess) is metallated as described in procedure a (see also Table 4). The red solution obtained is cooled to $-50°C$, after which freshly distilled trimethylchlorosilane (10% excess) is added. After the usual work-up the silylation product, b.p. 115 °C/15 mmHg, $n_D(20)$ 1.4762, is isolated in ~60% yield by careful distillation through a 30-cm Widmer column (b.p. limonene 60 °C/15 mmHg).

With BuLi·TMEDA in hexane [1] the metallation was found to proceed very sluggishly and the silylation product was obtained in a moderate yield in spite of the use of a 300% excess of limonene.

α-Pinene (compare Ref. [2])

a. Finely powdered t-BuOK (0.10 mol, 11.3 g) and freshly distilled α-pinene (0.4 mol, 54.4 g) are placed in the flask. A solution of 0.10 mol of BuLi in 66 ml of hexane is added over a few s to the stirred suspension. During this addition the temperature rises by 10 to 15 °C. After heating for 1 h at 50 °C, the brown suspension is cooled to below 0 °C and 50 ml of THF is added. The mixture is cooled to − 50 °C and 0.15 mol (16.3 g) of freshly distilled trimethylchlorosilane is added in one portion, whereupon the cooling bath is removed. After stirring for an additional 15 min at 0 to 10 °C, water is added and the silylation product, b.p. 95 °C/15 mmHg, is obtained in ~ 70% yield by careful fractionation (b.p. α-pinene 52 °C/18 mmHg) after the usual work-up. NMR spectroscopy indicates that only silylation in the methyl-group has occurred.

b. Following procedure b for the metallation, 1-heptene (see above) and α-pinene (300 mol% excess) can be converted into the silyl derivative with > 80% yields.

1,4-Cyclohexadiene

A solution of 0.10 mol of BuLi in 66 ml of hexane is placed in the flask and TMEDA (0.10 mol, 11.6 g) is added in one portion. The resulting solution is cooled to − 5 °C and cyclohexadiene (0.12 mol, 9.6 g, see exp. 14) is added in one portion. The yellow solution is stirred for 30 min at 0 °C and for an additional 30 min at + 5 °C. Pre-cooled (0 °C) THF (50 ml) is then added and, after cooling to − 40 °C, freshly distilled trimethylchlorosilane (0.11 mol, 11.9 g) is added in one portion. The cooling bath is removed and stirring is continued until the yellow colour has disappeared. A mixture of 25 g of 36% hydrochloric acid and 150 ml of water is then added with vigorous stirring, after which the product, b.p. 53 °C/12 mmHg, $n_D(20)$ 1.4759, is isolated in > 90% yield. GLC shows the presence of one component. The UV spectrum, showing one absorption at 198 nm (96% ethanol) or 203 nm (hexane), is in accordance with the structure of 3-trimethylsilyl-1,4-cyclohexadiene.

Allylbenzene

a. A mixture of 0.10 mol (11.8 g) of allylbenzene (exp. 15) and 80 ml of THF is cooled to − 25 °C, after which a solution of 0.105 mol of butyllithium in 69 ml of hexane is added over a few s from a syringe. The red solution is stirred for 10 min at 0 °C, after which the cooling bath is removed and the temperature allowed to rise to + 15 °C. The solution is then cooled to below − 40 °C and freshly distilled trimethylchlorosilane (0.11 mol, 11.9 g) is added in one portion. The resulting colourless reaction mixture is hydrolysed and the product (b.p. 110–115 °C/15 mmHg, yield > 90%) isolated in the usual way. It consists of a mixture of ~ 90% E-PhCH=CHCH$_2$SiMe$_3$ and ~ 10% of PhCH(SiMe$_3$)CH=CH$_2$.

b. A solution of 0.35 mol of sodamide in 300 ml of liquid ammonia is prepared in a 1-l three-necked round-bottomed flask, equipped with a dropping funnel, a mechanical stirrer and a vent (see Vol. 1, Chap. I). Allylbenzene (0.3 mol, 35.4 g) is added dropwise over 10 min with efficient stirring. Subsequently, ethyl bromide (0.45 mol, ~ 50 g) is added dropwise over 20 min to the red solution. After an additional 5 min, stirring is stopped and the ammonia allowed to evaporate. Water (200 ml) is then added with vigorous stirring and the mixture is extracted with Et$_2$O. After drying the organic solution over MgSO$_4$, the solvent is removed under reduced pressure and the remaining liquid carefully distilled through a 30-cm Widmer column. The fraction passing over between 65 and 76 °C/14 mmHg (yield ~ 70%) consists mainly (> 90%) of PhCH(C$_2$H$_5$)CH=CH$_2$. The minor component is PhCH=CHCH$_2$C$_2$H$_5$.

1,3-Pentadiene

A mixture of 0.10 mol (10.3 g) of t-BuOK, 70 ml of THF and 0.12 mol (8.2 g) of 1,3-pentadiene (commercially available) is cooled to − 90 °C (bath with liquid N$_2$). A solution of 0.10 mol of BuLi in 66 ml of hexane is added dropwise over 15 min, while keeping the mixture between − 75 and − 85 °C. A thick yellow suspension is formed. After the addition the cooling bath is removed and the temperature allowed to rise to 0 °C. The suspension is then cooled again to − 60 °C and 0.08 mol (13.2 g) of hexyl bromide is added in one portion. The reaction is very exothermic. After 5 min the cooling bath is removed and the temperature allowed to rise to 10 °C. Water is added with vigorous stirring and the product is isolated in the usual way (see above). Distillation gives a mixture of H$_2$C=CHCH(C$_6$H$_{13}$)CH=CH$_2$(~ 80%) and CH$_2$=CHCH=CHCH$_2$C$_6$H$_{13}$ (~ 20%), b.p. 60–72 °C/12 mmHg in ~ 75% yield.

Literature

1. Crawford RJ, Erman WF, Broaddus CD (1972) J Am Chem Soc 94:4298
2. Rauchschwalbe G, Schlosser M (1975) Helv Chim Acta 58:1094

7.8 Metallation of Cyclohexene

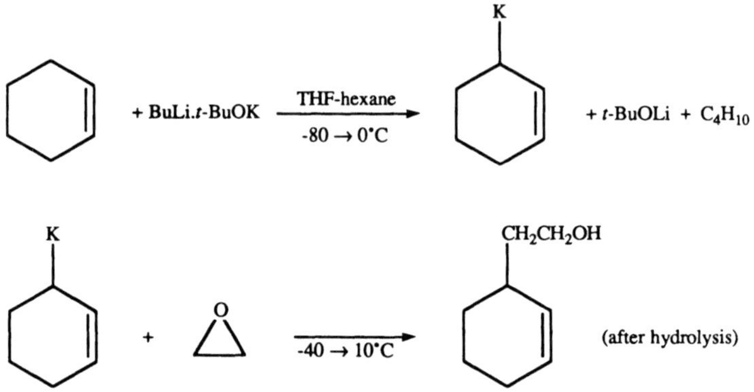

Apparatus: p. 24, 500 ml

Scale: 0.10 molar

Introduction

The allylic protons in cyclohexene are less acidic than those in aliphatic 1- and 2-alkenes. Prolonged (~ 24 h) treatment at $25\,^{\circ}$C of a $\sim 800\,$mol% excess of cyclohexene with a 1:1 molar mixture of *sec*.-BuLi and *t*-BuOK in isopentane, followed by addition of oxirane gave the corresponding alcohol in a good yield [1]. Our attempts to metallate cyclohexene with *n*-BuLi·TMEDA in hexane ($20\,^{\circ}$C or reflux) or with a 1:1:1 molar mixture of *n*-BuLi, *t*-BuOK, and TMEDA in hexane (-20 to $+10\,^{\circ}$C) gave poor results, presumably due to preferential attack of TMEDA by the base [2]. Reaction (at -80 to $0\,^{\circ}$C) of a five-fold molar excess of cyclohexene with *n*-BuLi·*t*-BuOK in a 1:1 mixture of THF and hexane, followed by reaction with oxirane gave the expected alcohol in only 35% yield. A much better result is obtained, however, if first the hexane is removed from the *n*-BuLi solution by evacuation. Yields in the region of 60% can be obtained, when cyclohexene is used in a four fold excess.

Procedure

A solution of 0.10 mol of *n*-BuLi in 66 ml of hexane is placed in the flask, which has been filled with inert gas. The flask is fitted with two stoppers and an evacuation device, after which it is placed in a water bath at $20\,^{\circ}$C. The greater part of the hexane is removed in a water-pump vacuum (a tube filled with KOH pellets is placed between the flask and the water aspirator). The flask is then placed in a cooling bath ($< -50\,^{\circ}$C) and inert gas is admitted. A mixture of 50 ml of THF and 0.5 mol (41 g) of cyclohexene (pre-cooled at $-40\,^{\circ}$C under an atmosphere of inert gas) is then added to the viscous solution of BuLi, after which the flask is equipped as indicated above. A solution of 0.10 mol of t-BuOK in 40 ml of THF is added with gentle stirring and cooling between -80 and $-85\,°C$ This addition takes ~ 15 min. The mixture is stirred for 2 h at -75 to $-80\,°C$, subsequently for 1 h at -65 to $-70\,°C$, then the cooling bath is removed and the temperature allowed to rise to $0\,°C$. The yellow suspension is then cooled again to $-50\,°C$, after which a mixture of 0.15 mol (6.6 g, excess) of oxirane and 30 ml of THF is added over 15 min, while the temperature is kept between -50 and $-30\,°C$. After an additional 30 min (at $-30\,°C$), the reaction mixture has become almost colourless. Water (100 ml) is then added with vigorous stirring. After separation of the layers, the aqueous layer is extracted six times with small portions of Et_2O. The combined organic solutions are dried over $MgSO_4$ and subsequently concentrated under reduced pressure. Distillation of the remaining liquid through a 20 cm Vigreux column gives the product, b.p. $65\,°C/1$ mmHg, $n_D(20)$ 1.4849, in 58–65% yield.

Literature

1. Hartmann J, Schlosser M (1976) Helv Chem Acta 59:453
2. Köhler FH, Hertkorn N, Blümel J (1987) Chem Ber 120:2081

7.9 Dimetallation of Isobutene

$$(CH_3)_2C{=}CH_2 + 2BuLi\cdot t\text{-BuOK} \xrightarrow[20\,°C]{\text{Hexane}} (KCH_2)_2C{=}CH_2 + 2t\text{-BuOLi}$$

$$(KCH_2)_2C{=}CH_2 + 2\ \text{oxirane} \xrightarrow[-30\to 0\,°C]{\text{THF-hexane}} (HOCH_2CH_2CH_2)_2C{=}CH_2$$

$$\text{(after hydrolysis)}$$

Apparatus: p. 24, 500 ml

Scale: 0.05 molar (isobutene)

Introduction

Prolonged interaction at room temperature between $BuLi\cdot TMEDA$ and an excess of isobutene affords isobutenyllithium, as appeared from the results of subsequent reactions with electrophilic reagents [1]. When the metallation is carried out with a 1:2 molar ratio of isobutene and $BuLi\cdot TMEDA$ in hexane, mainly dilithiated isobutene is formed [2]. Bates et al. [3] reacted isobutene with two mol equivalents of the Lochmann–Schlosser reagent in hexane and observed a much faster dimetallation. It should be pointed out that this dimetallation takes place under heterogeneous conditions and in the apolar, inert solvent hexane. Our attempts to achieve a more complete dimetallation by carrying out the reaction in the presence of TMEDA (BuLi, t-BuOK, and TMEDA, when mixed in a 1:1:1 molar ratio with hexane at − 20 °C give an almost homogeneous solution) gave only the mono-potassio compound. The extra equivalent of basic reagent presumably reacts with the TMEDA [5]. Bates et al. report that the dimetallation takes only about 10 min (at room temperature). This period seems too short to us since we observed evolution of heat over a period of ∼ 45 min after the addition of isobutene to the base. Another modification introduced by us concerns the use of an excess of BuLi and t-BuOK, which seems desirable for a more complete conversion into the dipotassio derivative. The excess of BuLi·t-BuOK can be easily destroyed by addition of THF. This reacts with the base to give ethene and $H_2C=CH-OK$ [4]. These compounds are not likely to interfere seriously in subsequent reactions with electrophiles. This also holds for the t-BuOLi in the reaction mixture of the metallation: the filtration procedure to remove the soluble t-BuOLi applied by Bates et al. is therefore not carried out.

Procedure

t-BuOK (0.12 mol, 13.6 g, finely powdered under inert gas) and hexane (70 ml) are placed in the flask. A solution of 0.12 mol of BuLi in 79 ml of hexane is added over a few min from a syringe with cooling between − 20 and 0 °C. The mixture is stirred for 30 min at 10 °C, then it is cooled to 0 °C and a mixture of 0.05 mol (2.8 g) of isobutene and 30 ml of hexane (cooled at − 20 °C) is added in one portion. The mixture is stirred for 1 h at 10–15 °C and subsequently for 1.5 h at + 20 °C. The yellowish-brown suspension is then cooled to 0 °C, 80 ml of THF is added and stirring at + 5 °C is continued for 20 min (during this period the excess of BuLi·t-BuOK reacts with the THF to give ethene and $H_2C=CH-OK$). After cooling the suspension to − 50 °C, a mixture of 0.15 mol (6.6 g, excess) of oxirane and 20 ml of THF is added in one portion. Ten minutes after this addition the cooling bath is removed and the temperature allowed to rise to 0 °C. A thin, almost white suspension is formed. The greater part of the solvent is then removed on the rotary evaporator. To the somewhat viscous residue 100 ml of Et_2O and 100 ml of water are successively added. After vigorous shaking and separation of the layers, ten extractions with small portions of Et_2O are carried out. The (unwashed) organic solutions are dried over $MgSO_4$ and subsequently concentrated in vacuo. The remaining liquid is distilled through a short column to give the diol $H_2C=C(CH_2CH_2CH_2OH)_2$, b.p. ∼ 120 °C/1 mm Hg, in ca. 60% yield. The smaller first fraction consists mainly of $H_2C=C(CH_3)CH_2CH_2CH_2OH$, formed from monometallated isobutene.

Literature

1. Akiyama S, Hooz J (1973) Tetrahedron Lett 4115
2. Klein J, Medlik A (1973) J Chem Soc, Chem Comm 275
3. Bates RB, Beavers WA, Gordon III B (1979) J Org Chem 44:3800
4. Lehmann R, Schlosser M (1984) Tetrahedron Lett 745
5. Köhler FH, Hertkorn N, Blümel J (1987) Chem Ber 120:2081

7.10 Metallation of Isoprene

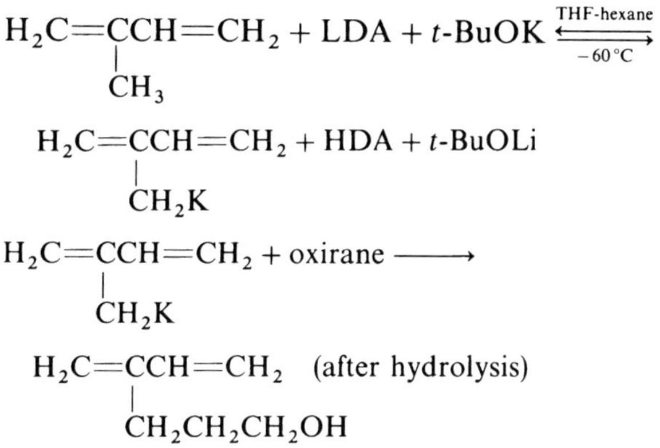

$$H_2C=CCH=CH_2 + LDA + t\text{-BuOK} \underset{-60\,°C}{\overset{\text{THF-hexane}}{\rightleftharpoons}}$$
$$\underset{CH_3}{|}$$

$$H_2C=CCH=CH_2 + HDA + t\text{-BuOLi}$$
$$\underset{CH_2K}{|}$$

$$H_2C=CCH=CH_2 + \text{oxirane} \longrightarrow$$
$$\underset{CH_2K}{|}$$

$$H_2C=CCH=CH_2 \quad \text{(after hydrolysis)}$$
$$\underset{CH_2CH_2CH_2OH}{|}$$

Apparatus: p. 24, 500 ml

Scale: 0.10 molar

Introduction (compare introduction of Exp. 11)

Isoprene undergoes a very ready polymerization under the influence of strongly basic reagents. With dialkylamides, however, the main reaction is deprotonation at the methyl group. It has been shown [1] that the position of the deprotonation equilibrium depends upon the pK value of the anion from which the amide is derived, and the alkali-metal counterion. Whereas in the case of *lithium* dialkyl-amides the equilibrium is almost completely on the side of the reactants, interaction between isoprene and the cesium and potassium amides (generated in situ from $LiNR_2$ and t-BuOCs or t-BuOK) gives rise to significant concentrations of the isoprene anion, as can be derived from the yields of the products obtained after addition of a higher alkyl bromide or oxirane. As expected from their greater thermodynamic basicities [2], potassium diisopropyl amide and potassium 2,2',6,6'-tetramethylpiperidide give higher yields than potassium diethylamide. The synthetic applicability of the metallation reaction is very limited: only the reactions with alkyl halides and oxirane give satisfactory yields, probably because these compounds react faster with the potassiated isoprene than with the potassium dialkylamide, thus causing the equilibrium to shift to the right side. The strong effect of the counterion upon the position of the metallation equilibrium is analogous to that observed in the metallation of aromatic hydrocarbons such as toluene [3] and α-methylstyrene (Exp. 11). Other examples are the reactions of 2-methylpyridine and methyl allyl sulfide $H_2C=CHCH_2SCH_3$ with alkali amides in liquid ammonia. With $LiNH_2$, these compounds give only very low concentrations of the anion, but interaction with potassium amide results in the formation of red solutions, which react with alkyl halides to give alkyl derivatives in good yields [4].

Procedure

t-BuOK (0.10 mol, 11.3 g) is dissolved in 80 ml of THF. Diisopropylamine (0.10 mol, 10.1 g) is added, after which the mixture is cooled to below $-60\,°C$. A solution of 0.10 mol of BuLi in 66 ml of hexane is then added over a few s from a syringe. The resulting solution is cooled to $-70\,°C$ and 0.15 mol (10.2 g) of freshly distilled isoprene is added over 10 min. A deeply red solution is formed immediately. After an additional 10 min (at -60 to $-70\,°C$), a mixture of 0.15 mol (6.6 g) of oxirane and 20 ml of THF is added in one portion. After stirring for 15 min at $\sim 50\,°C$, the cooling bath is removed and the temperature allowed to rise to $0\,°C$. Water (50 ml) is then added to the almost colourless solution, after which the aqueous layer is extracted six times with small portions of Et_2O. The combined organic solutions are dried (without previous washing) over $MgSO_4$ and subsequently concentrated under reduced pressure. Careful distillation through a 30-cm Vigreux column gives the alcohol, b.p. $77\,°C/10\,mmHg$, $n_D(20)$ 1.4787, in $\sim 55\%$ yield. When the metallation is carried out with potassium tetramethylpiperidide, yields of $\sim 70\%$ are obtained.

Literature

1. Klusener PAA, Hommes HH, Verkruijsse HD, Brandsma L (1985) J Chem Soc, Chem Comm 1677
2. Fraser RR, Mansour TS (1984) J Org Chem 49:3442
3. Ahlbrecht H, Schneider G (1986) Tetrahedron 42:4729
4. Unpublished results from the author's laboratory

7.11 Metallation of α-Methylstyrene

$$Ph—C(CH_3)=CH_2 + LDA·t\text{-BuOK} \xrightleftharpoons{\text{THF-hexane}}$$

$$Ph—C(CH_2K)=CH_2 + HDA + t\text{-BuOLi}$$

$$Ph—C(CH_2K)=CH_2 + \text{oxirane} \longrightarrow Ph—C(CH_2CH_2CH_2OH)=CH_2$$

$$\text{(after hydrolysis)}$$

Apparatus: p. 24, 500 ml

Scale: 0.10 molar

Introduction (compare introduction of preceding experiment).

Treatment of α-methylstyrene with BuLi in THF or Et_2O and with BuLi·TMEDA in hexane results only in 'anionic' polymerization. With a 1:1 molar mixture of BuLi and *t*-BuOK in THF and hexane at low temperatures polymerization and deprotonation at the methyl group take place to a comparable extent: after quenching with trimethylchlorosilane, the silyl derivative $PhC(CH_2SiMe_3)=CH_2$

can be isolated in ~ 30% yield. Although this result is somewhat better than that obtained with isoprene ($<$ 10% yields of quench product), it is not very interesting from a preparative point of view. It is obvious that for achieving deprotonation one has to use dialkylamides, preferably sterically hindered ones, such as diisopropylamide. As in the case of isoprene, the deprotonation equilibrium with LDA is completely on he left side (the solution remains colourless), but with a 1:1 molar mixture of LDA and t-BuOK a deeply red solution is formed. Quenching of this solution with Me_3SiCl gives the expected silyl derivative in a moderate yield. With oxirane, however, the corresponding alcohol is formed in a good yield. These experimental data are explainable by assuming that KDA ($=$ LDA $+ t$-BuOK) and α-methylstyrene are in equilibrium with α-potassiomethylstyrene and diisopropylamine. If this intermediate reacts faster with oxirane than does KDA, the equilibrium can shift to the right side and the hydroxyalkylation product is formed with a good yield. The considerably lower yield of the silyl derivative might be the consequence of a competing reaction of Me_3SiCl with KDA. Hydroxyalkylations with aldehydes and ketones generally are most successful with lithium compounds. However, addition of lithium bromide to the reaction mixture obtained from α-methylstyrene and KDA gives rise to a complete shift of the equilibrium to the left side (the solution becomes almost colourless). A similar phenomenon is observed in the case of isoprene. The derivatization possibilities of α-methylstyrene (and isoprene) via the metallic intermediates are therefore very limited.

Procedure

α-Methylstyrene (0.10 mol, 11.8 g) is added over a few min at $-70\,°C$ to the solution of potassium diisopropylamide prepared as described in Exp. 10. After 10 min a mixture of 0.15 mol (6.6 g) of oxirane and 20 ml of THF is added in one portion to the deep-red solution. The cooling bath is removed and the temperature allowed to rise to $0\,°C$. Dilute hydrochloric acid (4 N, 100 ml) is then added with vigorous stirring and, after separation of the layers, three extractions with Et_2O are carried out. The combined organic solutions are dried over $MgSO_4$ and subsequently concentrated in vacuo. Distillation through a short column gives the alcohol, b.p. $108\,°C/2\,mmHg$, $n_D(20)$ 1.5486, in 68% yield. The small first fraction is α-methylstyrene.

7.12 Metallation of Indene

Subsequent reactions with n-butyl bromide and trimethylchlorosilane.

Apparatus: p. 24, 500 ml, the dropping funnel is replaced by a rubber septum

Scale: 0.10 molar

Introduction

Owing to the possibility of aromatic π-delocalization in the anion, indene has a much lower pK value (~ 21) then aliphatic and cycloaliphatic olefins such as hexene and cyclohexadiene. The relatively high thermodynamic acidity is reflected in the ease with which indene is metallated. The reaction with BuLi in mixtures of THF and hexane at temperatures between 0 and $-50\,°C$ is extremely fast. In the more polar liquid ammonia, indene and lithium amide form a solution of indenyllithium. 1-Substituted indenes readily undergo base-catalyzed conversion to the 3-isomers. This isomerization can be induced by the metallated indene under the conditions of functionalization. (Inversed) addition of the metallated indene to an excess of the electrophilic reagent in a solvent of moderate polarity can help to prevent the isomerization. Swedish chemists [1,2] prepared a number of 1-alkylindenes using Et_2O as a solvent. Even the reaction with t-butyl halide gave 1-t-butylindene in a reasonable yield (a SET process rather than the usual S_N2-mechanism may be involved). A drawback of their procedure is the long reaction time (several hours at room temperature). In liquid ammonia the alkylations are undoubtedly extremely fast, but under these polar conditions a partial or even complete subsequent isomerization to 3-alkylindenes may be expected. In THF the alkylation of indenyllithium with primary alkyl bromides is much faster than in Et_2O. Using a 100% molar excess of alkyl bromide and applying the inversed-order addition technique, the formation of 3-alkylindene can be prevented. 1-(Trimethylsilyl)indene can be prepared by a similar procedure. Significant amounts of the 3-isomer and 1,3-bis(trimethylsilyl)indene are formed when the conventional procedure (addition of Me_3SiCl to indenyllithium) is followed.

Procedure

Indenyllithium

Indene (0.10 mol, 11.6 g) and THF (70 ml) are placed in the flask. The solution is cooled to $-40\,°C$ and a solution of 0.10 mol of butyllithium in 67 ml of hexane is added over a few min from a syringe. The cooling bath is removed and the orange solution can immediately be used for derivatization reactions.

Alkylation with butyl bromide

A mixture of 0.20 mol (27.4 g, 100% excess) of butyl bromide and 50 ml of THF is placed in the flask. The solution of indenyllithium is added over 10 min (by means of a syringe) to this mixture, while keeping the temperature between -40 and $-20\,°C$. After an additional 15 min the cooling bath is removed and the temperature allowed to rise to $0\,°C$. Water (100 ml) is then added after which the aqueous layer is extracted with Et$_2$O. The organic solution is dried over MgSO$_4$ and subsequently concentrated under reduced pressure. Careful distillation of the remaining liquid through a 30-cm Vigreux column gives 1-butylindene, b.p. 116 °C/12 mmHg, n$_D$(20) 1.5332, in ~ 75% yield. The small second fraction with b.p. 117–150 °C/12 mmHg, consists of 3-butylindene (up to 30%) and 1-butylindene.

Trimethylsilylation

A similar procedure is followed for the reaction of indenyllithium with trimethylchlorosilane (a 100% molar excess of Me$_3$SiCl is used). 1-(Trimethylsilyl)indene, b.p. 105 °C/12 mmHg, n$_D$(20) 1.5422, is obtained in ~ 70% by careful fractional distillation. The small fraction with b.p. 106–115 °C/12 mmHg contains ~ 20% of 3-(trimethylsilyl)indene and some of a bis(trimethylsilyl) derivative.

Literature

1. Cedheim L, Eberson L (1973) Synthesis 159
2. Meurling L (1974) Acta Chem Scand B 28:295 and Refs mentioned

7.13 Metallation of Cyclopentadiene

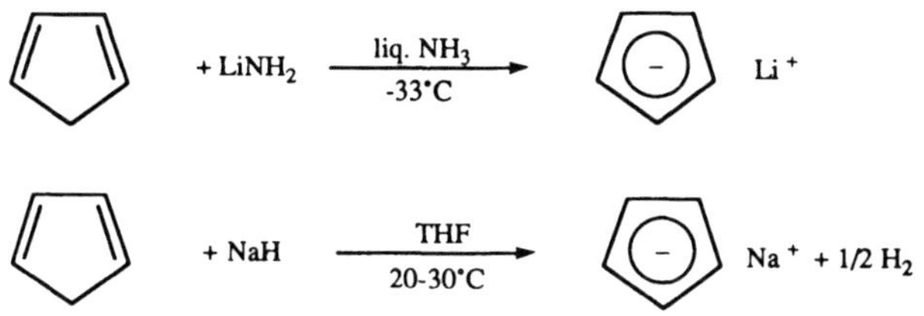

Derivatization reactions with n-hexyl bromide and trimethylchlorosilane.

Apparatus

For the reactions in liquid ammonia: 1-1 round-bottomed, three-necked (vertical necks) flask equipped with a dropping funnel, a mechanical stirrer and a vent; for the reaction in THF p. 24, Fig. 1, 11; the outlet is connected with a trap cooled in a bath with dry ice and acetone (< − 75 °C).

Scale: 0.3 to 0.5 molar

Introduction

The high acidity (pK ~ 15) of cyclopentadiene allows a variety of bases ranging from relatively weak to very strong to be applied for its metallation. The nature of the basic reagent is determined to a considerable extent by the functionalization reaction to be carried out. Reactions of cyclopentadienyllithium (CpLi) in Et_2O or THF with alkyl halides are extremely sluggish and give poor results, in part because of the slight solubility of CpLi in these solvents. In contrast, alkylations of CpLi in liquid ammonia, in which solvent its solubility is good, proceed smoothly. Reactions of organoalkali compounds with trimethylchlorosilane are usually not very demanding as to the solvent and the solubility of the alkali intermediate: our attempts to trimethylsilylate CpLi in Et_2O failed completely. On the other hand, CpNa, reasonably soluble in THF, is silylated smoothly. Reactions with aldehydes and ketones are not very demanding with respect to the solubility of the organometallic intermediate and are generally fast.

Another feature of cyclopentadiene chemistry is the strong tendency of Cp-derivatives to isomerize under the conditions of their preparation (compare the situation with indene derivatives). This reaction, in which the 5-substituted derivative is converted into the thermodynamically more stable 1-isomer, can be catalyzed by the alkali cyclopentadienylide. The easy isomerization is a consequence of the high acidity of the methine protons in the derivatives.

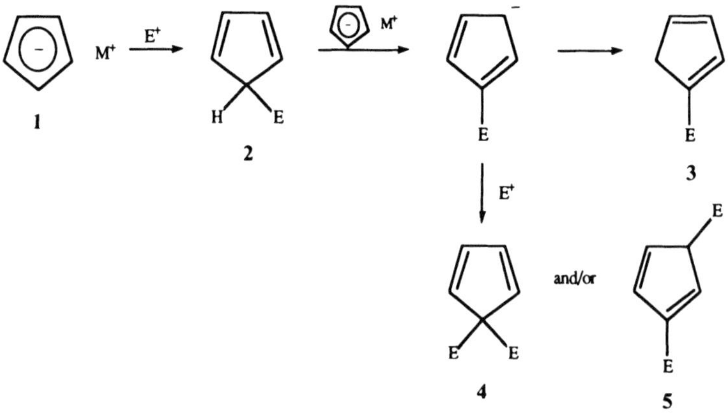

The isomerization may be accompanied by the introduction of a second group E as shown in the scheme. It does not seem easy to prevent or avoid these subsequent processes.

Procedure

Alkylation in liquid ammonia

Freshly prepared (by 'cracking' of the dimer) and redistilled cyclopentadiene (0.35 mol, 23.1 g) is added dropwise over 10 min to a vigorously stirred suspension of lithium amide in 350 ml of liquid ammonia prepared from 0.30 mol (2.1 g) of Li [1]. Losses of the volatile hydrocarbon due to sweeping along with the ammonia vapour can be avoided by occasionally cooling the reaction flask in a bath with dry ice and acetone or liquid nitrogen so that the ammonia is kept below its boiling point. To the grey solution hexyl bromide (0.30 mol, 49.5 g) is added in one portion (without using a cooling bath). The mixture from which salt begins to separate after a few minutes, is vigorously stirred for 1 h, then the remaining ammonia is removed by placing the flask in a water bath at 40 °C. Water (300 ml) is added to the remaining salt mass, after which three extractions with Et_2O are carried out. The organic solutions are dried over $MgSO_4$ and, after removing the solvent in vacuo, the remaining liquid is carefully distilled. 1-Hexyl-1,3-cyclopentadiene (3, E = C_6H_{13}), b.p. 80 °C/15 mmHg, formed during the reaction in ammonia from the initial alkylation product (2, E = C_6H_{13}) is obtained in ~ 65% yield. A considerable residue (4 or 5) is left behind.

Trimethylsilylation in THF

THF (350 ml) and sodium hydride (protected with paraffin oil; an amount corresponding to 0.55 mol) are placed in the flask (previous removal of the oil is not carried out). Freshly distilled cyclopentadiene (0.60 mol, 39.6 g, precooled below − 20 °C) is added dropwise over 1 h while cooling the flask in a bath at ~ 10 °C. After completion of the addition, the liquid in the cold trap (a mixture of THF and cyclopentadiene) is added to the reaction mixture and stirring is continued for an additional 30 min without external cooling while introducing nitrogen at a moderate rate (note 1). Freshly distilled trimethylchlorosilane (54.0 g, 0.50 mol) is added in five equal portions to the turbid solution over 15 min, while keeping the temperature between 15 and 25 °C. After an additional half hour water (~ 500 ml) is added with vigorous stirring. After separation of the layers, the organic solution is washed seven times with 150-ml portions of water in order to remove as much of the THF as possible. The combined aqueous layers are extracted twice with 75-ml portions of pentane. The combined pentane extracts are washed four times with water and subsequently combined with the first organic layer. After drying over $MgSO_4$, the greater part of the solvent is quickly distilled off at atmospheric pressure through a 40-cm Vigreux column keeping the bath temperature below 90 °C. After cooling to room temperature, the remaining liquid is subjected to a careful distillation through a 40-cm Widmer column. The fraction with b.p. 40–45 °C/12 mmHg, $n_D(20)$ 1,4635, consists of the 5-isomer (2, E = $SiMe_3$, main component ~ 85%) and the 1- and 2-isomer (~ 15%) in accordance with literature [2]. Our yield (~ 40%) is much lower than that reported in Ref. [2] (60%). At 90–95 °C/12 mmHg a considerable amount of a high-boiling fraction (probably a bis(trimethylsilyl)derivative, not reported in Ref. [2]) is collected.

Notes

1. CpNa is extremely sensitive to oxygen. Contact with air gives rise to the formation of a brown to black solution.

Literature

1. Brandsma L, Verkruijsse HD (1987) Preparative Polar Organometallic Chemistry, Vol 1, Chap I, Springer-Verlag
2. Abel EW, Dunster MO (1971) J Organometal Chem 33:161

7.14 Preparation of 1,4-Cyclohexadiene

$$\text{benzene} + 2\,\text{Li} + 2\,t\text{-BuOH} \xrightarrow[-33^\circ\text{C}]{\text{liq. NH}_3} \text{1,4-cyclohexadiene} + 2\,t\text{-BuOLi}$$

Apparatus: 3-l round-bottomed, three-necked flask, equipped with a stopper (which is temporarily replaced with a powder funnel during the addition of lithium), an efficient mechanical stirrer and a vent.

Scale: 1.0 molar

Procedure

Anhydrous liquid ammonia (1.5 l, drawn from a cylinder) is placed in the flask (for working with this solvent, see Vol. 1 and Brandsma L, Preparative Acetylenic Chemistry, Elsevier, 1988). Thiophene-free benzene (1.0 mol, 78 g) and t-butyl-alcohol (2.2 mol, 163 g) are subsequently added as a mixture. Lithium (2.0 mol, 14.0 g) is added in pieces of ~0.5 g over a period of 1.5 h. No pre-cut lithium should be used, since the cutting surface is immediately covered by a thin layer of oxide and nitride, making dissolution more difficult: the piece of lithium is kept above the powder funnel and is subsequently cut with a pair of scissors. If only thin (~2 mm) lithium wire is available, an excess of ~0.2 g has to be used to compensate for losses due to reaction of the relatively larger surface with oxygen, nitrogen, and moisture. The volume of the reaction mixture is kept at ~1.5 l. Losses of ammonia due to evaporation can be minimized by surrounding the flask by cotton wool, so that only a small volume of ammonia has to be added after 45 min. At the end an additional amount of ~1.5 g of metal is added and stirring is continued for 45 min. After this period the solution should be still deep-blue. High-boiling (b.p. > 170 °C/760 mmHg) petroleum ether (500 ml) is then cautiously poured into the flask. The mixture is then poured (over ~5 min) onto 2.5 kg of finely crushed ice which is contained (in two equal portions) in two large conical or round-bottomed flasks. After the ice has melted (some warming may be necessary) and the layers are separated, the aqueous layer is extracted three times with 100-ml portions of petroleum ether. The combined organic solutions are washed seven times with cold (0 °C) 4 M hydrochloric acid in order to remove the t-butylalcohol. After drying over the minimal amount of MgSO$_4$ (this salt is added in small portions with shaking until it remains in suspension), the solution is transferred into a 3–1 distillation flask. This is connected to a 40-cm Vigreux column, condenser and single receiver (cooled in a bath at < − 50 °C). The system is evacuated (water aspirator) and the flask gradually heated until the petroleum ether begins to reflux in the head of the column. Redistillation of the contents of the receiver gives cyclohexadiene (a trace of benzene may be present), b.p. 82 °C/760 mmHg, $n_D(20)$ 1.4746, in ~80% yield.

7.15 Allylbenzene

$$PhMgBr + H_2C{=}CHCH_2Br \xrightarrow[\text{reflux}]{Et_2O} PhCH_2CH{=}CH_2 + MgBr_2$$

Apparatus: 1-1, three-necked, round-bottomed flask, equipped with a dropping funnel combined with a gas inlet (for introducing inert gas), a mechanical stirrer and a reflux condenser.

Scale: 0.3 molar

Procedure

A solution of phenylmagnesium bromide is made in the usual way from 0.35 mol of bromobenzene, 0.5 mol (excess) of magnesium turnings and 250 ml of Et_2O. The solution is transferred (under inert gas) into the reaction flask, which previously has been filled with inert gas. The solution is brought to gentle reflux, after which the heating bath is removed and the addition of allyl bromide (0.30 mol, 36.3 g) is started. The addittion is carried out at a rate such that gentle reflux is maintained. When, after the addition (taking 15 to 20 min) refluxing has stopped, the reaction mixture is cooled to $\sim 0\,°C$ by placing the flask in a bath with ice and ice water. Copper(I) bromide or chloride (0.5 g) is added. The cooling bath is removed and stirring is continued for half an hour (this catalyst is used to ensure a complete conversion). The reaction mixture is cautiously poured into a solution of 0.5 mol of ammonium chloride in 300 ml of ice water (the NH_4Cl solution should never be added to the reaction mixture since much heat is evolved). After shaking, the layers are separated and the aqueous layer is extracted once with Et_2O. The organic solution is dried over $MgSO_4$ and subsequently concentrated in vacuo. The remaining liquid is distilled through a 30-cm Vigreux column to give allylbenzene, b.p. $45\,°C/12\,mmHg$, $n_D(20)$ 1.5118, in greater than 80% yield.

8 Selected Procedures from Literature

Murphy WS, Hamrick PJ, Hauser CR (1973) Org Synth Coll 5:523
See also: Hauser CR, Hamrick PJ (1957) J Am Chem Soc 79:3142

$$Ph_2CH_2 \xrightarrow[\text{2. } C_4H_9Br]{\text{1. } NaNH_2,\ liq.\ NH_3} Ph_2CHC_4H_9$$

Eisch JJ (1981) Organometallic Syntheses 2:91, Academic Press; Eisch JJ, Jacobs AM (1963) J Org Chem 28:2145:

$$H_2C{=}CHCH_2OPh + 2Li \xrightarrow{THF} H_2C{=}CHCH_2Li + PhOLi$$

Eisch JJ (1981) Organometallic Syntheses 2:98, Academic Press; Gilman H, Gaj BJ (1963) J Org Chem
    28:1725

$$Ph_3CH + BuLi \xrightarrow{\text{THF} + Et_2O} Ph_3CLi \xrightarrow[\text{2. H}^+, H_2O]{\text{1. } CO_2} Ph_3C-COOH$$

Trost BM, Chan DMT, Nanninga TN (1984) Org Synth 62:58:

$$H_2C{=}C(CH_3)CH_2OH \xrightarrow[\text{2.2Me}_3\text{SiCl}]{\text{1. 2BuLi.TMEDA}} H_2C{=}C(CH_2SiMe_3)CH_2OSiMe_3$$

Chapter III
Metallation of Saturated Sulfur Compounds

1 Introduction

In a preliminary communication Arens and the brothers Fröling reported on the metallation of formaldehyde diethylthioacetal, $H_2C(SC_2H_5)_2$, with sodamide in liquid ammonia and subsequent reaction of the obtained solution with alkyl halides [1]. In a full paper [2] Fröling and Arens described further examples of this metallation-functionalization. In the case of the oxygen analogue no conversion was observed (very recently [5]) Shiner et al. described the generation of the unstable lithiated (oxygen) acetals by reductive elimination of PhS from $(RO)_2CHSPh$ at very low temperature). The use of butyllithium (which in the mean time had become commercially available) in a mixture of THF and hexane allowed Corey and Seebach to fully explore this synthetic principle [3]. Since the thioacetal moiety can be converted into a carbonyl function by a number of methods [4], the overall operation of functionalizing anions $^-C(SR)_2$ with various electrophiles and converting the products $E-C(SR)_2$ into $E-C=O$ may be represented as nucleophilic acylation.

The greater part of this chapter is devoted to the metallation of S,S-acetals and some related compounds and subsequent reactions of the organometallic intermediates with a number of electrophilic reagents. Representative conditions for the metallation of saturated sulfur compounds are given in Table 5.

Literature

1. Arens JF, Fröling M, Fröling A (1959) Recl Trav Chim Pays-Bas 78:663
2. Fröling A, Arens JF (1962) *ibid* 81:1009
3. Seebach D (1969) Synthesis 17
4. Gröbel BT, Seebach D (1977) Synthesis 357
5. Shiner CS, Tsunoda T, Goodman BA, Ingham S, Lee S, Vorndam PE (1989) J Am Chem Soc 111:1381

2 Substrates and Metallation Conditions

2.1 S,S-Acetals

Most of the syntheses with S,S-acetals published after the first review of Seebach [1] have been carried out with 1,3-dithianes as starting compound. Butyllithium has been commonly used as metallating agent in a solvent mixture of THF and hexane.

In many cases the more easily and more cheaply available open-chain S,S-acetals, such as $H_2C(SCH_3)_2$ and $H_2C(SC_2H_5)_2$, can be used as starting compounds instead of 1,3-dithianes. We found that the abstractions of a methylene proton from 1,3-dithiane and $H_2C(SCH_3)_2$ or $H_2C(SC_2H_5)_2$ by BuLi and the subsequent reactions with butyl bromide proceeded with comparable rates. Differences arise, however, when the *alkyl* derivatives have to be lithiated. Whereas the lithiation of 2-alkyl-1,3-dithianes by BuLi seems to proceed cleanly [1], the open-chain derivatives

Table 5. Preparative reaction conditions for the metallation of saturated sulfur compounds

Substrate	Base–solvent[a]	Temp. range[b] (°C)	Additional time[c] (min)
CH_3SCH_3	BuLi–TMEDA-hexane	$+20 \rightarrow +40$	30
CH_3SPh	BuLi–TMEDA-hexane	$+20 \rightarrow +40$	15
CH_3OCH_2SPh	BuLi–THF-hexane	$-60 \rightarrow -40$	45
2-Alkyl-1,3-dithiane	BuLi–THF-hexane	$-20 \rightarrow +10$	45
$H_2C(SCH_3)_2$	BuLi–THF-hexane	$-30 \rightarrow +10$	20
	BuLi–TMEDA-hexane	$-20 \rightarrow +10$	10
	$KNH_2–NH_3(liq)$	-33	0
1,3-dithiane	BuLi–THF-hexane	$-30 \rightarrow +10$	20
$H_2C(SPh)_2$	BuLi–THF-hexane	$-30 \rightarrow -10$	30
$Me_3SiCH(SCH_3)_2$	BuLi–THF-hexane	$-40 \rightarrow 0$	20
$PhCH(SCH_3)_2$	BuLi–THF-hexane	$-50 \rightarrow 0$	5
$PhCH_2SiMe_3$	BuLi–THF-hexane	$0 \rightarrow 30$	10
$H_2C(SCH_3)_2$	$MNH_2–NH_3(liq)$	-33	0[d]
Alkyl-$CH(SCH_3)_2$	$KNH_2–NH_3(liq)$	-33	0[e]
$PhCH(SCH_3)_2$	$MNH_2–NH_3(liq)$	-33	0
	BuLi–THF-hexane	$-40 \rightarrow 0$	20
$HC(SCH_3)_3$	BuLi–THF-hexane	$-80 \rightarrow -60$	15[f]
$H_2C(S(=O)Et)(SEt)$	BuLi–THF-hexane	$-70 \rightarrow -30$	0
	LDA–THF-hexane	$-50 \rightarrow 0$	0
	$LiNH_2–NH_3(liq)$	-33	0
$CH_3S(=O)CH_3$	BuLi–THF-hexane	$-80 \rightarrow -20$	0
	LDA–THF-hexane	$-10 \rightarrow 0$	30
	$KNH_2–NH_3(liq)$	-33	0
$CH_3S(=O)Ph$	LDA–THF-hexane	$-20 \rightarrow +10$	30
	$NaNH_2–NH_3(liq)$	-33	0
CH_3SO_2Ph	BuLi–THF-hexane	$-70 \rightarrow -30$	0[g]

[a] Initial concentration 0.5 to 0.8 mol/l; scale 0.05 to 0.1 molar; substrate and base are used in equivalent amounts.
[b] After addition of substrate to base over a few minutes with cooling in a bath with dry ice- acetone or liquid N_2, the bath is removed and the temperature allowed to rise to the highest value indicated.
[c] Time for $\sim > 95\%$ completion of the reaction at highest temperature of the range indicated.
[d] M = Li, Na, K; for M = Li, the metallation equilibrium is strongly on the side of the reactants.
[e] Even with KNH_2 as the base, the ionization is probably incomplete.
[f] At higher temperatures elimination of $LiSCH_3$ occurs.
[g] The solution of the base is added to a mixture of the sulfone and THF.

$RCH(SR')_2$ (R = alkyl, R' = alkyl or phenyl) undergo a side-reaction whereby evil-smelling products, possibly $RCH(SH)SR'$ (after hydrolysis), are formed [2].

With alkali amides in liquid ammonia, abstraction of a methylene or the methyne proton is the only process observed:

$$RCH(SR')_2 + MNH_2 \rightleftharpoons RC(M)(SR')_2 + NH_3 \quad M = Li, Na, K$$

Since the thermodynamic acidities of NH_3 and thioacetals with R = H or alkyl and R' = alkyl are comparable [3], the position of the deprotonation equilibrium depends strongly upon the nature of the counterion. With KNH_2 as metallating agent the concentration of the S,S-acetal 'anion' is considerably higher than in the case of $LiNH_2$. This is illustrated in a convincing way by adding $CH_3CH(SC_2H_5)_2$ to equivalent amounts of alkali amide in liquid ammonia and subsequently introducing an alkyl halide RX: in the case of $LiNH_2$ no $CH_3C(R)(SC_2H_5)_2$ is formed, but with KNH_2 the yield of the dialkyl derivative is ca. 60% [4].

The thioacetal derived from benzenethiol, $H_2C(SPh)_2$, is expected to be somewhat more acidic than the alkyl derivatives $H_2C(SR')_2$. Replacement of a methylene proton in $H_2C(SR')_2$ by a phenyl group gives rise to a considerable increase of the (thermodynamic) acidity of the remaining methyne proton.

LDA in general seems not suitable as a reagent for the lithiation of $RCH(SR')_2$ (deprotonation too slow and incomplete) unless R is a strongly activating group, e.g., $COCH_3$ or Ph. Potassium diisopropylamide (in THF-hexane mixtures), prepared by combining equivalent amounts of LDA and t-BuOK [5-8], presumably will cause deprotonation of alkyl-$CH(SR')_2$ at a sufficiently high rate affording alkyl-$C(K)(SR')_2$ in a high equilibrium concentration. The complications observed with BuLi will probably not occur. The solutions of the potassiated S,S-acetals thus obtained seem particularly suitable for reactions with alkyl halides and epoxides.

The metallation of (cyclic) formaldehyde S,S-acetals with BuLi in THF-hexane mixtures (concentration of reagents 0.5–0.8 mol/l) is usually carried out at temperatures in the region of −20 °C and is completed within 1 to 2 hours. Alkyl derivatives $RCH(SR')_2$ (R = alkyl) react more slowly. The reaction times can be shortened by carrying out the metallations at 0–10 °C (above 10 °C THF is attacked by BuLi) or by allowing the temperature to rise to 0 or 10 °C. Most lithiated S,S-acetals have a reasonable thermal stability in the temperature region 0 to 20 °C.

Literature

1. Seebach D (1969) Synthesis 17
2. Brandsma L, Andringa H (unpublished observation)
3. Streitwieser Jr A, Ewing SP (1975) J Am Chem Soc 97:190; Bordwell FG, Matthews WS, Vanier NR
   ibid, p 442
4. Fröling A, Arens JF (1962) Recl Trav Chim Pays-Bas 81:1009
5. Renger B, Hügel H, Wykypiel W, Seebach D (1978) Chem Ber 111:2630
6. Lochmann L, Trekoval J (1979) J Organom Chem 179:123
7. Ahlbrecht H, Schneider G (1986) Tetrahedron 42:4729
8. Klusener PAA, Hommes HH, Verkruijsse HD, Brandsma L (1985) J Chem Soc Chem Comm
   1677

2.2 Methoxymethyl Phenyl Sulfide

Methoxymethyl phenylsulfide, $PhSCH_2OCH_3$, is readily accessible by reaction of benzenethiolate, PhS^-, with the chloromethyl ether, $ClCH_2OCH_3$, or by BF_3-assisted displacement of a CH_3O-group in $H_2C(OCH_3)_2$ by PhSH [3]. Lithiation of $PhSCH_2OCH_3$ is effected by treatment at $\sim -70\,°C$ with sec-BuLi.TMEDA in a THF-hexane mixture or with n-BuLi at $\sim -40\,°C$ in the same solvent mixture [1–3]. The reason for using the more strongly basic sec-BuLi at very low temperatures by some workers possibly is the low stability of $PhSCH(Li)OCH_3$, compared to lithiated S,S-acetals.

$$PhSCH_2OCH_3 \xrightarrow{\leq\,-40\,°C} PhSCH(Li)OCH_3 \xrightarrow{>\,-20\,°C} decomposition$$

We found that $PhSCH_2OCH_3$ can be lithiated completely within 45 min at temperatures between -40 and $-45\,°C$ using n-BuLi in THF and hexane. Subsequent functionalizations gave high yields. When during the lithiation or derivatization reaction the temperature was allowed to rise above $-20\,°C$, however, reduced or low yields of impure products were obtained. Since reactions with most electrophiles can be completed within 1 to 2 hours at temperatures below $-40\,°C$, the lower stability of $PhSCH(Li)OCH_3$ is not a serious drawback in syntheses performed on a small or moderate scale. Problems may arise if first an alkyl chain is introduced and the methyne proton in the product $PhSCH(Alkyl)OCH_3$ is to be replaced by lithium. 1,3-Oxathiane has been incidentally used as a substrate for lithiation-functionalization reactions [4].

Literature

1. de Groot Ae, Jansen BJM (1981) Tetrahedron Lett 887
2. Trost BM, Miller CH (1975) J Am Chem Soc 97:7182
3. Hackett S, Livinghouse T (1986) J Org Chem 51:879
4. Fuji K, Ueda M, Sumi K, Fujita E (1981) Tetrahedron Lett 2005

2.3 Ethylthiomethyl Ethyl Sulfoxide

Since the methylene protons in the compounds $H_2C(SOCH_3)SCH_3$ and $H_2C(SOC_2H_5)SC_2H_5$ are considerably more acidic than those in the corresponding dialkyl thioacetals, the less strongly basic LDA can be used instead of butyllithium for the lithiation. During treatment of $H_2C(SOCH_3)SCH_3$ with base, competitive metallation in the sulfoxide methyl group occurs, therefore the ethyl compound is preferred in syntheses [1].

$$H_2C(SOCH_3)SCH_3 \longrightarrow \underset{\underset{SCH_3}{|}}{LiCH(SOCH_3)} + \underset{\underset{SCH_3 \text{ (minor)}}{|}}{H_2C-S(=O)-CH_2Li}$$

$$H_2C(SOC_2H_5)SC_2H_5 \longrightarrow \underset{\underset{SC_2H_5}{|}}{LiCH(SOC_2H_5)}$$

In their review, Gröbel and Seebach [2] give a summary of the relative advantages and disadvantages of the use of S,S-acetals and the corresponding S-oxides in organic syntheses.

Literature

1. Richman JE, Herrmann JL, Schlessinger RH (1973) Tetrahedron Lett 3267
2. Gröbel BT, Seebach D (1977) Synthesis 357

2.4 Orthothioformates

Orthothioformates $HC(SR)_3$ (R = alkyl or aryl) are smoothly metallated by alkali amides in liquid ammonia and by butyllithium in THF and hexane. The anionic intermediates decompose readily to give thiolates RS^- and carbenes $(RS)_2C$: [1]. Further reaction of the latter species (presumably with $(RS)_3C^-$) gives rise to the formation of dimers $(RS)_2C=C(SR)_2$:

$$(RS)_3CH \longrightarrow (RS)_3C^-M^+ \longrightarrow (RS)_2C: + RS^-M^+$$
$$\downarrow (RS)_3C^-M^+$$
$$(RS)_2C=C(SR)_2 + RS^-M^+$$

These carbenoid reactions occur already at temperatures in the region of $-40\,°C$. Good results in functionalization reactions are attainable only with electrophiles that prefer $(RS)_3C^-$ over RS^-. Generation of the carbenoids at very low temperatures is also an important condition [2].

Literature

1. Wildschut GA, Bos HJT, Brandsma L, Arens JF (1967) Monatsh Chem 98:1043
2. Seebach D (1972) Chem Ber 105:487

2.5 Dialkyl Sulfides and Alkyl Aryl Sulfides

Dimethyl sulfide and methyl phenyl sulfide are readily and cleanly lithiated at ambient or slightly elevated temperatures by the BuLi·TMEDA-hexane reagent and by the 1:1 molar combination of BuLi and 1,4-diazabicyclo[2,2,2]octane (DABCO) in hexane [1–3]. Dimethyl sulfide can also be metallated by a 1:1 molar mixture of BuLi and t-BuOK in THF-hexane at $\sim -80\,°C$ [4].

$$CH_3SR \longrightarrow LiCH_2SR \qquad R = CH_3 \text{ or phenyl}$$

Butyllithium (in THF) does not react with dimethyl sulfide, while with methyl phenyl sulfide metallation occurs in the ring as well as in the methyl group [5]. Homologues of CH_3SPh, RCH_2SPh, have been specifically lithiated in the side chain using t-BuLi in a THF-hexane HMPT-mixture [6]. With BuLi·TMEDA ring-metallation and elimination reactions take place.

$$RCH_2SPh \longrightarrow RCH(Li)SPh$$

Another manner to obtain α-lithioalkyl aryl sulfides is to react alkyllithium, $R'Li$, with vinyl phenyl sulfide in diethyl ether at $0\,^\circ C$ [7]:

$$H_2C=CHSPh + R'Li \longrightarrow R'CH_2CH(Li)SPh$$

Cyclopropyl sulfides are metallated much more easily than dialkyl sulfides. Exclusive ring metallation is observed upon interacting cyclopropyl-SPh with n-BuLi in THF at $0\,^\circ C$ [8]. Using a 1:1 molar mixture of BuLi and t-BuOK in THF and hexane $(-70\,^\circ C)$ we obtained comparable amounts of ring-metallated product and cyclopropyl-$S-CH_2K$ [4].

Benzyl methyl sulfide, $PhCH_2SCH_3$, is preferentially metallated on the methylene carbon atom [9].

Literature

1. Peterson DJ (1967) J Org Chem 32:1717
2. Corey EJ, Seebach D (1966) J Org Chem 31:4097
3. Cabiddu S, Floris C, Melis S, Sotgiu F (1984) Phosphorus and Sulfur 19:61; see also (1986) Tetrahedron Lett 4625
4. Brandsma L (unpublished)
5. Shirley DA, Reeves BJ (1969) J Organom Chem 16:1
6. Dolak TM, Bryson TA (1977) Tetrahedron Lett 1961
7. Ager DJ (1981) Tetrahedron Lett 22:587
8. Trost BM, Keeley D, Bogdanowicz MJ (1973) J Am Chem Soc 95:3068
9. Brandsma L (unpublished)

2.6 Dialkyl and Alkyl Aryl Sulfoxides and Sulfones

Dimethyl sulfoxide reacts at temperatures in the range of $80\,^\circ C$ with sodium hydride to give a solution of 'dimsylsodium', $NaCH_2SOCH_3$ [1]. For most derivatization reactions this solution is not very attractive, since the work-up involves separation of the desired product (in many cases having a good water-solubility) from large amounts of DMSO.

Addition of a solution of BuLi in hexane to a strongly cooled solution of DMSO in THF gives a white suspension of dimsyllithium, with which further synthetic operations may be carried out [2]:

$$CH_3S(=O)CH_3 + BuLi \xrightarrow{-70\,^\circ C} LiCH_2(S=O)CH_3$$

LDA also gives satisfactory results.

For the metallation of aryl alkyl sulfoxides, BuLi is not a suitable base, since the main reaction is cleavage of the S-aryl bond [3]. The desired metallation in the alkyl group can be effected with LDA and with sodamide or potassium amide in liquid ammonia [4, 5]:

$$CH_3S(=O)Ph + LDA \xrightarrow{\text{THF-hexane}} LiCH_2S(=O)Ph$$

$$CH_3S(=O)Ph + KNH_2 \xrightarrow{\text{liq. NH}_3} KCH_2S(=O)Ph$$

Dialkyl and alkyl arylsulfones are much more acidic than the corresponding sulfoxides. Interaction with alkali amides in liquid ammonia or LDA in THF-hexane mixtures gives rise to a complete metallation, but also butyllithium can be successfully used [2], e.g.:

$$CH_3SO_2Ph \xrightarrow[\text{solvent}]{\text{MNH}_2, \text{ LDA or BuLi}} MCH_2SO_2Ph \qquad (M = Li, Na, K)$$

Methyl phenyl sulfone is readily dilithiated to give the geminal dilithio derivative [6].

Literature

1. Corey EJ, Chaykovsky M (1962) J Am Chem Soc 84:866; (1965) 87:1345
2. Field L (1972) Synthesis 115:122; (1978) Synthesis 732
3. Lockard JP, Schroeck CW, Johnson CR (1973) Synthesis 485
4. Tsuchihashi G, Iriuchijima S, Maniwa K (1973) Tetrahedron Lett 3389
5. Brandsma L (unpublished)
6. Bongini A, Savoia D, Umani-Ronchi U (1976) J Organometal Chem 112:1

3 Experiments

All temperatures are internal, unless indicated otherwise.
All reactions are carried out in an atmosphere of inert gas.
For general instructions concerning handling organolithium reagents, drying solvents etc., see Vol. 1.

3.1 Lithiation of Formaldehyde Dimethylthioacetal by BuLi in THF and Hexane

$$H_2C(SCH_3)_2 + BuLi \xrightarrow[-40 \rightarrow 0°C]{\text{THF-hexane}} LiCH(SCH_3)_2 + C_4H_{10}$$

Apparatus: p. 24, Fig. 1; 11

Scale: 0.10 molar

Introduction

This experiment gives typical reaction conditions for the lithiation of formaldehyde dialkylthioacetals $H_2C(SR)_2$. The rates of metallation of $H_2C(SCH_3)_2$, $H_2C(SC_2H_5)_2$, and 1,3-dithiane are comparable, while aryl compounds $H_2C(S\text{-Aryl})_2$ are expected to be lithiated somewhat faster. The original literature [1] prescribes temperatures between -20 and $-40\,°C$ for the lithiation of 1,3-dithiane with BuLi in THF-hexane mixtures. The reaction times can be shortened considerably by allowing the temperature to rise to 0 or $+10\,°C$ (initial concentrations of reactants 0.5 to 0.8 mol/l). The lithiated S,S-acetals are fairly stable at temperatures up to about $20\,°C$ as suggested by the results of our alkylation experiments. It is advisable, however, to keep the temperature below $10\,°C$ during the lithiation, since at higher temperatures THF may be attacked by unreacted BuLi or by lithiated thioacetal (for the lithiation with BuLi.TMEDA in hexane see exp. 2). In diethyl ether the lithiations of $H_2C(SCH_3)_2$ and $H_2C(SC_2H_5)_2$ with BuLi are considerably slower than in THF, but in most syntheses there will be no reason for preferring Et_2O over THF. Treatment of the open-chain S,S-acetals, alkyl-$CH(SR)_2$, with BuLi in THF-hexane or with BuLi·TMEDA in hexane gives rise to a seriously competing side-reaction, possibly attack by the base on the group R (compare Ref. [1]). These problems do not seem to occur with 2-alkyl-1,3-*dithianes*.

Procedure (compare Refs. [1–4])

A mixture of 0.10 mol (10.8 g) of formaldehyde dimethylthioacetal (see Exp. 21) and 80 ml of THF is placed in the flask. After cooling to $-50\,°C$, a solution of 0.105 mol of butyllithium in 68 ml of hexane is added over a few seconds by means of a syringe, while keeping the temperature below $-30\,°C$. Immediately after this addition, the cooling bath is removed and the temperature allowed to rise to $+5\,°C$ (temporary cooling may be necessary). After being stirred for an additional 15 min at $+5\,°C$, the almost clear and slightly yellow solution can be used for further synthetic operations (note 1). Functionalization reactions should be carried out not later than 1 h after the metallation, during which period the solution of $LiCH(SCH_3)_2$ should be kept below $10\,°C$.

Literature

1. Seebach D (1969) Synthesis 17. Gives a general procedure for the lithiation of 1,3-dithiane and derivatives
2. Seebach D, Beck AK (1971) Organic Syntheses 51:76. In this procedure 1,3-dithiane is lithiated with BuLi in THF-hexane at -10 to $-20\,°C$. The initial concentration of 1,3-dithiane is about four times as low as that of $H_2C(SCH_3)_2$, the reaction time for the lithiation of 1,3-dithiane is 2 h
3. Seebach D, Erickson BW, Singh G (1966) J Org Chem 31:4303. 2-Phenyl-1,3-dithiane is lithiated with BuLi in THF-hexane at $-50\,°C$. The reaction time is 6 hours. Regarding the activating influence of the phenyl group, this time seems rather long, but it should be borne in mind that in the next step lithium has to be replaced by deuterium. For obtaining very pure 2-deuterio-2-phenyl-1,3-dithiane by deuteration, a conversion into the lithio compound of say 97% is not satisfactory. The long reaction

time at $-50\,°C$ in the Organic Synthesis procedure is therefore rational. It could perhaps have been shortened by allowing the temperature to rise to $0\,°C$.

4. Seebach D, Corey EJ (1975) J Org Chem 40:231

Notes

1. From the data in Ref. [4] it can be derived that most S,S-acetals $H_2C(SR')_2$ and $RCH(SR')_2((SR')_2 = $ open-chain or 1,3-dithianyl) can be metallated 'completely' ($\geq 95\%$) within 2 h at $-20\,°C$, using BuLi in THF-hexane (1:1) and concentrations of reagents of 0.3 mol/l or higher.

3.2 Lithiation of Formaldehyde Dimethylthioacetal with BuLi·TMEDA in Hexane

$$H_2C(SCH_3)_2 + BuLi \xrightarrow[-20\to0°C]{\text{TMEDA-hexane}} LiCH(SCH_3)_2 \cdot TMEDA\downarrow + C_4H_{10}$$

Apparatus: p. 24, Fig. 1; 11

Scale: 0.10 molar

Introduction: see Exp. 1

Procedure

A solution of 0.105 mol of BuLi in 69 ml of hexane is placed in the flask. TMEDA (0.12 mol, 13.9 g, dried by distillation from $LiAlH_4$ in a partial vacuum) is added over a few minutes. An additional amount of 50 ml of hexane or pentane is added, after which the solution is cooled to $-40\,°C$. The S,S-acetal (0.10 mol, 10.8 g, see Exp. 21) is then introduced over a few s, after which the cooling bath is removed. The temperature rises to $0-10\,°C$ within a few minutes. After an additional 5 min (at $\sim 10\,°C$) the desired co-solvent (Et_2O or THF, usually 80 ml) is added to the thick suspension with cooling below $0\,°C$. The resulting suspension (in the case of Et_2O) or solution (in the case of THF) should be immediately cooled to below $-40\,°C$, after which the reaction with the desired compound is carried out.

3.3 Reaction of Lithiated Bis(methylthio)methane with Alkyl Halides

$$LiCH(SCH_3)_2 + RBr \xrightarrow{\text{THF-hexane}} RCH(SCH_3)_2 + LiBr$$

$$(R = C_4H_9, CH_2\text{-}CH\!=\!CH_2, PhCH_2)$$

Apparatus: p. 24, Fig. 1; 11

Scale: 0.10 molar

Introduction

This experiment gives typical conditions for the reaction of lithiated $H_2C(SCH_3)_2$ with alkylating agents. Under similar conditions lithiated aromatic compounds, Aryl-CH_2Li, lithiated pyridines, e.g., 2- and 4-picolyllithium, and lithiated alkenes, e.g., $H_2C=CHCH_2Li$, react much faster, while lithio-imines such as $LiCH_2CH=N-t-C_4H_9$ are less reactive towards alkyl halides than lithiated S,S-acetals. We presume that the reactivities of $LiCH(SCH_3)_2$, $LiCH(SPh)_2$, 2-lithio-1,3-dithiane, and several lithiated derivatives $RC(Li)(SR')_2$ with $R = $ alkyl or phenyl towards the halogen compounds are comparable.

In Seebach's reviews [1, 2] several examples of successful alkylation reactions with lithiated S,S-acetals (mainly 1,3-dithianes) are mentioned. These data and our experimental results gave rise to the following summary of the reactivities of lithiated S,S-acetals in alkylations and the result:

1. Ethyl bromide and higher homologues react completely within 1 h with lithiated S,S-acetals (concentration 0.5 to 0.8 mol/l in THF-hexane 1:1 v/v) in the temperature range $-30°$ to $0°C$. CH_3Br, CH_3I and higher homologues react smoothly at lower temperatures. Yields with simple primary alkyl bromides or iodides are generally high.
2. Lithiated S,S-acetals in THF-hexane mixtures discriminate excellently between chlorides and iodides (and probably also bromides), allowing very selective couplings with $Cl(CH_2)_nI$ or $Cl(CH_2)_nBr$ ($n \geqslant 3$).
3. Allyl bromide and other compounds with the system $C=C-C-Br$ react very smoothly and cleanly at temperatures in the region of $-80°C$. Benzyl bromide reacts easily at about $-80°C$ to give the expected product in an excellent yield. The *chloride* reacts smoothly between -60 and $-80°C$, but the product ($\sim 50\%$ yield) is very impure.
4. The bromoacetal $BrCH_2CH(OC_2H_5)_2$ and $LiCH(SCH_3)_2$ or $LiCH(SC_2H_5)_2$ gave the expected products $(C_2H_5O)_2CHCH_2CH(SR)_2$ in moderate yields (compare Ref. [1]) probably due to elimination of HBr and C_2H_5OH. Bromoacetaldehyde diethylacetal is much less reactive than primary alkyl bromide in nucleophilic displacements. In order to achieve a rate of conversion comparable with that in the case of n-butyl bromide, the reaction of $LiCH(SR)_2$ with the bromoacetal had to be carried out at 10 to 20°C [3].
5. Secondary alkyl bromides (e.g., isopropyl bromide) react less readily and higher temperatures are required. To avoid any decomposition of the lithiated thioacetal or its reaction with THF, the alkylations are usually carried out with the iodides [1]. Cyclopentyl bromide (reaction carried out at $\sim +10°C$) and $LiCH(SCH_3)_2$ gave the expected product in a moderate yield [3]. With cyclohexyl bromide the predominant reaction is dehydrobromination [1].

Procedure (compare Ref. [4])

A solution of 0.10 mol of $LiCH(SCH_3)_2$ (see Exp. 1) is cooled to $-40°C$. Butyl bromide (0.12 mol, 16.4 g) is added over 5 min, while maintaining a temperature between -40 and $-30\,°C$. After the addition, the cooling bath is removed. The temperature rises to between 10 and 15 °C within 15 min. The conversion is completed by warming the solution for an additional 15 min at 30 °C. After cooling to 0 °C, 100 ml of ice water is added with vigorous stirring. The layers are separated and the aqueous layer is extracted twice with small portions of ether. The organic solutions are dried over $MgSO_4$ and subsequently concentrated under reduced pressure. Distillation of the remaining liquid through a 30-cm Vigreux column gives $C_4H_9CH(SCH_3)_2$, b.p. 85 °C/12 mmHg, $n_D(20)$ 1.5040, in an excellent yield.

Allyl bromide (0.12 mol, 14.5 g) is added over 10 min with cooling between -70 and $-40\,°C$. After the addition the cooling bath is removed and stirring is continued for 15 min. The product $H_2C{=}CHCH_2CH(SCH_3)_2$, b.p. 77 °C/12 mmHg, $n_D(20)$ 1.5275, is obtained in an excellent yield.

Benzyl bromide (0.10 mol, 17.1 g) is added over 10 min with cooling between -70 and $-80\,°C$. The cooling bath is removed and the temperature allowed to rise to $-30\,°C$. Distillation after the usual work-up gives $PhCH_2CH(SCH_3)_2$, b.p. 153 °C/12 mmHg (20-cm Vigreux column), $n_D(20)$ 1.5860, in $>80\%$ yield.

Literature

1. Seebach D (1969) Synthesis 17
2. Gröbel BT, Seebach D (1977) Synthesis 357
3. Brandsma L (unpublished)
4. Seebach D, Corey EJ (1975) J Org Chem 40:231. A general procedure for the alkylation of 2-lithiated 1,3-dithianes is given. In our opinion the reaction times mentioned in this paper can be shortened considerably; compare Seebach D, Beck AK (1971) Org Syntheses 51:76

3.4 Hydroxyalkylation of Lithiated Bis(methylthio)methane with Epoxides

$LiCH(SCH_3)_2$ + $RCH{-}CH_2$ (epoxide) $\xrightarrow{\text{THF-hexane}}$ $(CH_3S)_2CHCH_2CH(R)(OH)$

(after hydrolysis)

R=H, CH₃, Ph

$LiCH(SCH_3)_2$ + (cyclohexene oxide) $\xrightarrow{\text{THF-hexane}}$ (product with OH and $CH(SCH_3)_2$)

(after hydrolysis)

Apparatus: p. 24, Fig. 1; 11

Scale: 0.10 molar

Procedure (compare Refs. [1–3])

A solution of 0.1 mol of lithiated bis(methylthio)methane (see Exp. 1) is cooled to $-40\,°C$ and a mixture of 0.15 mol (6.6 g) of oxirane and 20 ml of THF is added over

10 min while keeping the temperature between -40 and $-30\,°C$. After the addition the cooling bath is removed. The temperature may rise to ca. 10 or $15\,°C$. The almost colourless solution is then concentrated under reduced pressure (rotary evaporator). Et_2O (100 ml) is added after which the mixture is vigorously shaken with 200 ml of ice water. The organic layer and two extracts with Et_2O are dried (without washing) over $MgSO_4$ and subsequently concentrated under reduced pressure. Distillation through a 10 to 20 cm Vigreux column gives the alcohol (R = H), b.p. $\sim 90\,°C/1$ mmHg, $n_D(20)$ 1.5396, in an excellent yield. (Compare Exp. 17)

Epoxypropane (50% excess) and epoxystyrene (10% excess) are added between -20 and $-30\,°C$, after which the temperature is allowed to rise. The hydroxyalkylations with these epoxides seem somewhat less fast than in the case of oxirane. Stirring at 10 to $15\,°C$ is continued for 20 min, after which the same procedure of working up as described above is followed. The alcohols R = CH_3, b.p. $\sim 80\,°C/0.5$ mmHg, $n_D(20)$ 1.5297, and R = Ph, b.p. $\sim 140\,°C/0.05$ mmHg, $n_D(20)$ 1.5853, are obtained in > 80% yields.

Epoxycyclohexane (20% excess) is added in one portion at $-20\,°C$, after which the temperature is allowed to rise (without cooling bath) to $15\,°C$. Above $0\,°C$ the heating effect is easily observable. After an additional 45 min (at 15 to $20\,°C$) the product (b.p. $\sim 120\,°C/0.5$ mmHg, $n_D(20)$ 1.5519, yield > 90%) is isolated as described above.

Note

The ^1H- and ^{13}C-NMR spectra of the products obtained from propylene oxide, styrene oxide, and cyclohexene oxide show two CH_3S-signals.

Literature

1. Seebach D, Corey EJ (1975) J Org Chem 40:231. The reactions with epoxides are described as being 'very slow' and reaction times prescribed for β-hydroxyalkylations are of the order of 1 week at -5 or $-20\,°C$
2. Seebach D, Jones NR, Corey EJ (1968) J Org Chem 33:300; reaction times 16 to 46 hours
3. Jones JB, Grayshan R (1972) Can J Chem 50:810, 1407, 1414

3.5 Reaction of Lithiated 1,3-Dithiane with 1-Bromo-3-chloropropane and Ring Closure of the Coupling Product Under the Influence of Butyllithium

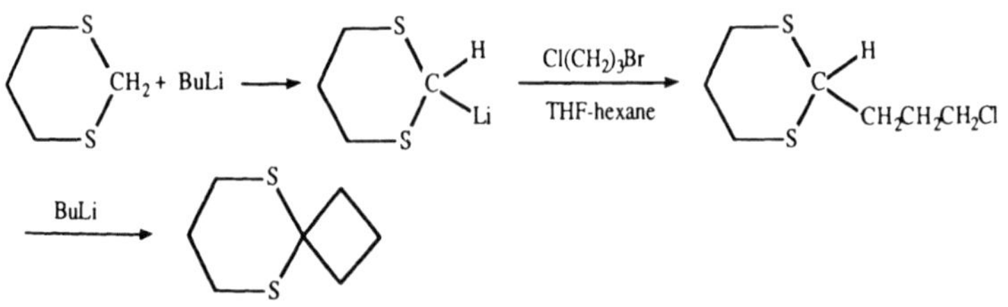

Apparatus: p. 24, Fig. 1; 11

Scale: 0.10 molar

Introduction

The synthesis of the spiro compound is based on the Organic Syntheses procedure in Ref. [1]. The modifications introduced by us concern mainly reaction temperatures and times. Cyclobutanone dimethylthioacetal can be prepared similarly with excellent results starting from bis(methylthio)methane [2].

Procedure

A solution of 0.105 mol of butyllithium in $\sim$ 70 ml of hexane is placed in the flask and 35 ml of THF is added with cooling below 0 °C. Subsequently, a mixture of 0.10 mol (12.0 g) 1,3-dithiane and 35 ml of THF is added in one portion at -40 °C. The cooling bath is removed and the temperature allowed to rise to $+10$ °C. After an additional 15 min at 10 °C the solution is cooled to -50 °C, 0.13 mol (20.4 g) of bromochloropropane is added over 5 min and the cooling bath is removed. Above -35 °C the heating effect becomes strong and the temperature may rise to 15 or 20 °C. The greater part of the THF and hexane is then removed under reduced pressure using a rotary evaporator and keeping the bath temperature below 40 °C. Ice water (200 ml) and Et_2O (100 ml) are then added with vigorous swirling. The aqueous layer is extracted once with Et_2O. The combined organic solutions are washed with water, dried over $MgSO_4$ and subsequently concentrated using the rotary evaporator. The excess of bromochloropropane is then removed at $p < 0.5$ mmHg, keeping the bath temperature below 50 °C. The residue is the almost pure 1,3-dithiane derivative, $n_D(20)$ 1.564, yield $\sim 98\%$. Prolonged standing at 20 °C gives a crystalline sulfonium salt.

The crude product is dissolved in 70 ml of THF and the solution cooled to -20 °C. A solution of 0.11 mol of butyllithium in 72 ml of hexane is added dropwise over 15 min while maintaining a temperature of -20 to -25 °C. After the addition the cooling bath is removed and the temperature allowed to rise to 10 °C. After an additional 30 min (at 20 °C) ice water (200 ml) is added with vigorous stirring. Separation of the layers, extraction of the aqueous layer with Et_2O (twice), drying over $MgSO_4$, and concentration of the organic solution in vacuo gives the crude spirocyclic compound. Distillation through a 10- to 20-cm Vigreux column gives the pure product, b.p. $\sim 55-60$ °C/0.5 mmHg, $n_D(20)$ 1.5691, in 82% yield.

Literature

1. Seebach D, Beck AK (1971) Organic syntheses 51:76
2. Brandsma L (unpublished)

3.6 Hydroxymethylation of Bis(methylthio)methane with Paraformaldehyde

$$\text{LiCH(SCH}_3)_2 + (\text{CH}_2\text{O})_n \xrightarrow[20 \to 35\,^\circ\text{C}]{\text{THF-hexane}} \text{HOCH}_2\text{CH(SCH}_3)_2$$

(after hydrolysis)

Apparatus: p. 24, Fig. 1; 11

Scale: 0.10 molar

Introduction

The reaction of lithiated S,S-acetals with aldehydes and ketones is a well-established reaction and needs no exemplification by an experimental procedure. The reactions with most of the aldehydes and ketones are extremely rapid, even at temperatures in the region of $-80\,^\circ\text{C}$. A number of examples have been mentioned in the reviews of Seebach [1, 2].

In this experiment it is shown that a CH_2OH group can be successfully introduced by reaction of the organolithium compound with (dry) para-formaldehyde [3]. For obtaining good yields of the corresponding alcohol, it appeared to be necessary to use a large excess (at least 400 mol%) of the polymeric aldehyde.

Procedure

To a solution of 0.10 mol of the lithiated S,S-acetal in THF and hexane (Exp. 1) is added in one portion at $10\,^\circ\text{C}$ 15 g ($\to 0.5$ mol of the monomer) finely powdered and dry (note 1) paraformaldehyde (note 2), At $15–18\,^\circ\text{C}$ a strongly exothermic reaction starts and in spite of cooling in a bath at $-78\,^\circ\text{C}$ (CO_2-acetone) the temperature may rise to between 30 and $40\,^\circ\text{C}$. As soon as the exothermic reaction is over, the cooling bath is removed and the mixture is stirred for an additional half hour at 25 to $30\,^\circ\text{C}$. The greater part of the THF and hexane is then removed under reduced pressure using a rotary evaporator and a heating bath at 35 to $40\,^\circ\text{C}$. Et_2O (150 ml) and ice water (150 ml) are successively added with vigorous stirring or swirling. After separation of the layers, the aqueous layer is extracted four times with small portion of Et_2O. The unwashed organic solution is dried well over MgSO_4 and subsequently concentrated in a water aspirator vacuum. Careful distillation of the remaining liquid through a 30-cm Vigreux column gives, after a small first fraction of $\text{H}_2\text{C(SCH}_3)_2$ the alcohol, b.p. $100–105\,^\circ\text{C}/12\,\text{mmHg}$, $n_D(20)$ 1.5520, in $\sim 80\%$ yield.

Notes

1. A larger amount of the powder is heated for half an hour in a vacuum of 10–20 mmHg using a rotary evaporator and a bath heated at 60–70 °C. It is advisable to put the powder in a relatively big flask (0.5 to 1 liter). A trap cooled at -78 °C is placed between the evaporator and the water aspirator.
2. The polymer is added through a powder funnel which replaces the dropping funnel.

Literature

1. Seebach D (1969) Synthesis 17
2. Gröbel BT, Seebach D (1977) Synthesis 357
3. Schaap A, Brandsma L, Arens JF (1967) Recl Trav Chim Pays-Bas 86:393

3.7 Reaction of Lithiated Bis(methylthio)methane with Dimethylformamide and Subsequent Acid Hydrolysis

$$LiCH(SCH_3)_2 + HC(=O)N(CH_3)_2 \xrightarrow[-100 \to -60\,°C]{THF\text{-}hexane}$$

$$(CH_3S)_2CHCH(OLi)NMe_2 \xrightarrow[H_2O]{H^+} (CH_3S)_2CHCH=O$$

Apparatus: p. 24, Fig. 1; 11

Scale: 0.10 molar

Introduction

A general method for the introduction of formyl groups consists of adding N,N-dimethylformamide (other N,N-disubstituted formamides may also be used) to a solution of an organoalkali or Grignard derivative in Et_2O or THF and subsequently treating the reaction mixture with dilute acid. A number of examples of this reaction with alkali compounds have been described in our previous books [1–3]. Grignard derivatives generally react less easily. With strongly basic organoalkali compounds the addition proceeds rapidly at very low temperatures, but with acetylides, $RC\equiv CM$, being weaker bases, no conversion is observed below -30 °C. The yields of this formylation reaction are generally good to excellent, provided that experimentation is carried out in a correct manner. Particularly during the hydrolysing procedure things may go wrong. Especially sensitive aldehydes may react with the dialkylamine (addition to systems $C=C-CH(=O)$) or may undergo aldol-condensation (systems $A\text{-}CH_2CH(=O)$) in which A is an electron-withdrawing group), if during the hydrolysis the medium is allowed to become alkaline. It seems, therefore, in many cases essential to use an excess of acid and to stir efficiently in order to avoid local concentrations of dialkylamine and alkali hydroxide. The literature [4–6] reports some examples of formylation of lithiated 1,3-dithianes. The reaction is carried out with an excess of DMF at temperatures between − 20 and − 5 °C. After several hours ('one night') the reaction mixture is hydrolysed. During this time the initial adduct rearranges to form an enolate:

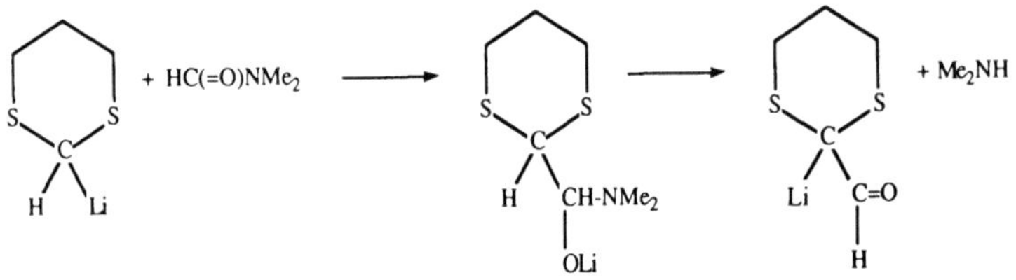

Apparently dimethylamine is eliminated under these conditions. The hydrolysis only serves to liberate the aldehyde from the enolate.

Procedure

A solution of 0.10 mol of lithiated bis(methylthio)methane in 70 ml of THF and 68 ml of hexane (Exp. 1) is cooled to − 100 °C (bath with liquid N$_2$). Dimethylformamide (0.15 mol, 11.0 g, p.a. quality) is added over a few s with vigorous stirring. After the addition the cooling bath is removed and the temperature allowed to rise to − 60 °C. The reaction is almost instantaneous. The clear solution is transferred (by means of a syringe or slowly poured through a funnel) into another 1-1 flask, containing a vigorously stirred mixture of 25 g of 36% hydrochloric acid and 200 ml of ice water. This operation takes ∼ 5 min. Care should be taken that the solution of the adduct does not run down the glass wall before it comes into contact with the acidic solution: this may give rise to temporary local high concentrations of base. After this addition a concentrated aqueous solution of potassium carbonate is introduced dropwise (while continuing stirring) until the aqueous layer has become pH 6. The layers are separated and the aqueous layer extracted three times with Et$_2$O. After drying the unwashed organic solution over MgSO$_4$, the solvent is removed under reduced pressure. Distillation of the remaining yellow liquid through a 20-cm Vigreux column gives the aldehyde, b.p. 90 °C/18 mmHg, n$_D$(20) 1.5449, in ca. 80% yield.

Literature

1. Brandsma L, Verkruijsse HD (1981) Synthesis of acetylenes, allenes and cumulenes, Elsevier
2. Brandsma L (1988) Preparative acetylenic chemistry, Revised Edition, Elsevier
3. Brandsma L, Verkruijsse HD (1987) Preparative polar organometallic chemistry, Vol. 1 Springer-Verlag
4. Seebach D, Corey EJ (1975) J Org Chem 40:231
5. Meyers AI, Strickland RC (1972) J Org Chem 87:2582
6. Wilson SR, Mathew J (1980) Synthesis 625

3.8 Reaction of Lithiated Bis(methylthio)methane with Carbon Dioxide

$$LiCH(SCH_3)_2 + CO_2 \xrightarrow[< -80\,°C]{THF\text{-}hexane} LiOOCCH(SCH_3)_2$$

$$\xrightarrow[H_2O]{H^+} HOOCCH(SCH_3)_2$$

Apparatus

1-l four-necked, round-bottomed flask equipped with a long gas-inlet tube, a thermometer and an efficient mechanical stirrer. The fourth neck is left open.

Scale: 0.10 molar

Introduction

2-Substituted 1,3-dithianes have been carboxylated with good yields. The usual procedure involves pouring the solution of the lithiated dithiane onto powdered dry ice, covered with Et_2O (the addition may also be carried out by means of a syringe) [1, 2]. With the unsubstituted dithiane the yields were moderate, due to occurrence of a very fast trans-metallation [2].

$$(RS)_2CH\text{-}Li \xrightarrow{CO_2} (RS)_2CH-COOLi \xrightarrow{1} (RS)_2C=C(OLi)_2$$
$$(I)$$

About 50% of the starting compound is recovered after the work-up. We have experienced the same problem, when a solution of lithiated bis(methylthio)methane in THF and hexane was added at temperatures above $-70\,°C$ to THF through which gaseous CO_2 was bubbled. The acid was obtained in only 45% yield. In the procedure described below (experimentally not easy) a very high concentration of CO_2 is attained by rapidly introducing carbon dioxide in THF cooled below $-80\,°C$. Other strongly basic organometallics can be successfully carboxylated in a similar way. A necessary condition is that they have a good solubility in THF: suspensions can much less easily be added from a syringe, while a slight solubility of the organometallic compound may result in an incomplete carboxylation.

Procedure

Dry THF (150 ml) is placed in the flask. Carbon dioxide is introduced from a cylinder, while the THF is vigorously stirred and the flask is cooled in a bath with liquid nitrogen. When the temperature has dropped to below $-80\,°C$, the flow of carbon dioxide is increased to about 3 l/min (a flow-meter should be used). The temperature of the solution is kept closely around $-90\,°C$. When after 3 min, a large amount of CO_2 has dissolved, the addition of the solution (0.10 mol) of the lithiated

S,S-acetal, by means of a syringe, is started. This solution has been brought at ca. $-30\,°C$. The addition (through the open neck) is carried out in such a way that the end of the needle of the syringe is held very near to the end of the inlet tube. During this operation, which takes $\sim 5\,min$, a CO_2-flow of ~ 3 l/min is maintained. Efficient cooling (with vigorous stirring) is necessary to keep the temperature between -80 and $-90\,°C$. A white suspension is formed. After the addition the flow of CO_2 is stopped, the cooling bath removed and the temperature of the suspension allowed to rise to $0\,°C$. Water (150 ml) is then added with vigorous stirring. The layers are separated and the organic layer shaken twice with 30-ml portions of water. These aqueous layers are combined with the first one, after which three extractions with Et_2O are carried out (to remove dissolved neutral compounds). Subsequently, cold 5 M hydrochloric acid is added to reach pH 2. The carboxylic acid is isolated by extracting five times with Et_2O, drying the unwashed extracts over $MgSO_4$ and removing the solvent in vacuo. The last traces of Et_2O and some water are then removed by heating the residue during 45 min at $50\,°C$ in a vacuum of $< 0.5\,mmHg$. The white solid obtained weighs 12.5 g, corresponding with a yield of 82% (compare Ref. [3]). The ^{1}H-NMR spectrum (CD_3COCD_3 as solvent) shows the product to be pure. It showed the following signals: 2.2 ppm (SCH_3), 4.5 ppm (HC), 10.8 ppm (COOH). Crystallization from a 3:1 mixture of hexane and Et_2O (at $-20\,°C$) gives a m.p. of $77.0–78.4\,°C$.

Literature

1. Seebach D (1969) Synthesis 17
2. Seebach D, Corey EJ (1975) J Org Chem 40:231
3. Schmidt M, Weissflog E (1976) Z Naturforsch 31b:136

3.9 Reaction of Lithiated Bis(methylthio)methane with Dimethyl Disulfide and Trimethylchlorosilane

$$(CH_3S)_2CHLi + CH_3SSCH_3 \xrightarrow[-60\ to\ -80\,°C]{THF\text{-}hexane} (CH_3S)_3CH + CH_3SLi$$

$$(CH_3S)_2CHLi + Me_3SiCl \xrightarrow[-60\ to\ -80\,°C]{THF\text{-}hexane} (CH_3S)_2CH—SiMe_3 + LiCl$$

Apparatus: p. 24, Fig. 1; 11

Scale: 0.10 molar

Introduction

Reactions of organoalkali compounds with disulfides and trialkylchlorosilane are often extremely fast. The usual procedure involves addition of the reagents to a cooled solution of the organoalkali intermediate in Et_2O or THF. In many cases the expected products are obtained in high yields [1]. The results may be poor, however, when the carbon atom bearing the metal is linked to one or two hydrogen atoms and one or two activating groups. The introduction of the RS or R_3Si group leads to an additional activation of the remaining proton or protons. The initial thiolation or silylation product may then be deprotonated by the unconverted organoalkali compound to give a new organometallic intermediate which may be sulfenylated or silylated to give the bis-thiolated or bis-silylated product, e.g.:

$$(CH_3S)_2CHLi \xrightarrow{CH_3SSCH_3} (CH_3S)_3CH \xrightarrow{(I)} (CH_3S)_3CLi$$

$$(I)$$

$$\xrightarrow{CH_3SSCH_3} (CH_3S)_4C$$

These subsequent reactions can be avoided by applying inversed-order addition, i.e., by adding the organometallic compound to the electrophilic reagent, the latter preferably used in excess. In this way excellent yields are obtained in the reactions of bis(methylthio)methyllithium with dimethyl disulfide and trimethylchlorosilane (compare Ref. [2]).

Procedure

In the flask is placed a mixture of 0.20 mol (18.8 g) of dimethyl disulfide or 0.20 mol (21.7 g) of freshly distilled (note 1) trimethylchlorosilane and 100 ml of Et_2O. The mixture is cooled to between -60 and $-80\,°C$ and a solution of 0.10 mol of bis(methylthio)methyllithium (see Exp. 1) is added over 10 min from the dropping funnel to the vigorously stirred mixture, while maintaining the temperature between -60 and $-80\,°C$ (occasional cooling in a bath with liquid N_2 permits this quick addition). After the addition the cooling bath is removed and the temperature allowed to rise to $-10\,°C$. Ice water (150 ml) is then added with vigorous stirring. The upper layer and one ethereal extract are dried over $MgSO_4$ and subsequently concentrated under reduced pressure. Distillation through a 20-cm Vigreux column gives $(CH_3S)_3CH$, b.p. $100\,°C/12$ mmHg, $n_D(20)$ 1.5749, and $(CH_3S)_2CHSiMe_3$, b.p. $80\,°C/12$ mmHg, $n_D(20)$ 1.5082, in greater than 90% yields.

Note

1. Trimethylchlorosilane from a bottle that has been frequently opened, may contain dissolved hydrochloric acid: this may destroy part of the organolithium compound.

Literature

1. Brandsma L, Verkruijsse HD (1987) Preparative Polar Organometallic Chemistry, Vol 1, Springer-Verlag
2. Gröbel BT, Seebach D (1977) Synthesis 379

3.10 Lithiation of Methoxymethyl Phenyl Sulfide
and Subsequent Reaction with Dimethylformamide

$$CH_3OCH_2SPh + BuLi \xrightarrow[-50°C]{THF\text{-}hexane} CH_3OCH(Li)SPh$$

$$CH_3OCH(Li)SPh + HC(=O)NMe_2 \xrightarrow[-70°C]{}$$

$$CH_3O-CHCH(OLi)NMe_2 \xrightarrow{H^+,H_2O} CH_3O-CH-CH(=O)$$
$$\quad\quad\quad\quad | \quad\quad\quad\quad\quad\quad\quad\quad\quad\quad\quad\quad | $$
$$\quad\quad\quad SPh \quad\quad\quad\quad\quad\quad\quad\quad\quad\quad SPh$$

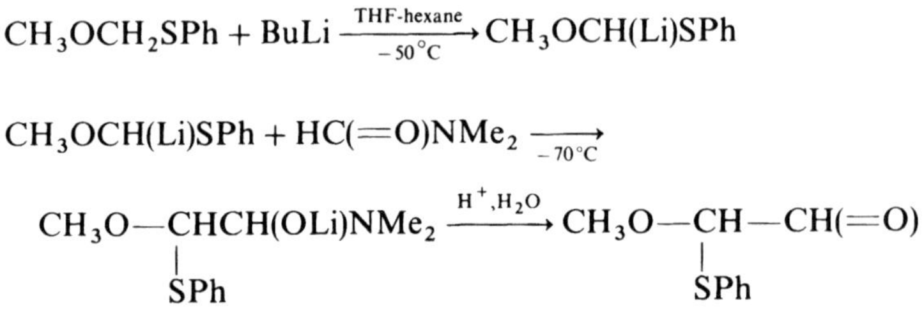

Apparatus: p. 24, Fig. 1; 11

Scale: 0.10 molar

Introduction

Whereas thioanisole, CH_3SPh, is reported to give a mixture of ring-metallated and side-chain-metallated products upon interaction with butyllithium, the O,S-acetal CH_3OCH_2SPh is lithiated only on the methylene carbon atom by BuLi in THF or sec-BuLi·TMEDA in THF [1–3]. Analogous metallations have been realized with 1,3-oxathiane [4]. The obtained lithium compounds have a limited thermal stability [5], so functionalizations with alkyl halides and epoxides, which are usually less fast than reactions with other 'electrophiles', give reduced yields. The O,S-acetal $PhSCH_2OCH_3$ can be lithiated in a reasonable time with butyllithium in a THF-hexane mixture but the temperature has to be kept below − 40 °C to prevent decomposition of the lithiated intermediate into PhSLi and other (unidentified) products [5].

The reaction of $PhSCH(Li)OCH_3$ with dimethylformamide proceeds almost instantaneously at temperatures in the region of − 80 °C and acid hydrolysis affords the formyl derivative in a high yield.

Procedure

A mixture of 0.10 mol (15.4 g) of methoxymethyl phenyl sulfide [2] and 80 ml of THF is placed in the flask and a solution of 0.11 mol of butyllithium in ~ 70 ml of hexane is added dropwise over 10 min, while keeping the temperature between − 50 and − 55 °C. After an additional period of 1.5 h at − 50 °C, the solution is cooled to − 70 °C and 0.15 mol (11.0 g) of dry DMF is added over a few s with vigorous stirring. The cooling bath is removed and the temperature allowed to rise to − 20 °C. The solution is then poured into a vigorously stirred mixture of 28 g of concentrated (36%) hydrochloric acid and 200 ml of ice water. The aqueous layer is brought to ~

pH 6 by dropwise addition of a concentrated aqueous solution of potassium carbonate. The organic layer and two ethereal extracts are dried over $MgSO_4$ and, after concentration in vacuo, the residue is distilled through a short Vigreux column to give the aldehyde, b.p. $\sim 100\,°C/0.5\,mmHg$, $n_D(20)$ 1.5687, in $\sim 80\%$ yield.

Addition of trimethylchlorosilane (0.13 mol, 14.1 g freshly distilled) at $-70\,°C$ in one portion to the solution of $CH_3OCH(Li)SPh$ and allowing the temperature to rise to $-20\,°C$ gave $CH_3OCH(SiMe_3)SPh$ in an excellent yield and with a satisfactory purity. The product was not distilled.

Literature

1. Trost BM, Miller CH (1975) J Am Chem Soc 97:7182
2. Hackett S, Livinghouse T (1986) J Org Chem 51:879
3. de Groot A, Jansen BJM (1981) Tetrahedron Lett 22:887
4. Fuji K, Ueda M, Sumi K, Fujita E (1981) Tetrahedron Lett 22:2005
5. Heus-Kloos YA, Brandsma L (unpublished observations)

3.11 Reaction of Lithiated Bis(methylthio)methane with Methyl Isothiocyanate and N,N-Dimethylcarbamoyl Chloride

$$(CH_3S)_2CHLi + CH_3N{=}C{=}S \xrightarrow[-80 \to -20\,°C]{THF\text{-}hexane} (CH_3S)_2CH{-}C({=}NCH_3)SLi$$

$$\xrightarrow{H^+, H_2O} (CH_3S)_2CH{-}C({=}S)NHCH_3$$

$$(CH_3S)_2CHLi + ClC({=}O)NMe_2 \xrightarrow[-70 \to -20\,°C]{THF\text{-}hexane} (CH_3S)_2CH{-}C({=}O)NMe_2$$

Apparatus: p. 24, Fig. 1; 500 ml

Scale: 0.05 molar

Introduction

Reactions of strongly basic organometallic compounds like $(CH_3S)_2CHLi$ with isothiocyanates and isocyanates are extremely fast even at very low temperatures. The adducts $(CH_3S)_2CHC(SLi){=}NCH_3$ and $(CH_3S)_2CHC(OLi){=}NCH_3$ still possess a rather acidic methyne proton, so there is some chance that they are deprotonated by unconverted $(CH_3S)_2CHLi$ if the heterocumulene is added slowly and at too high a temperature. Such processes result in lower yields of the N-substituted thioamides and amides since part of the lithiated S,S-acetal is used in these subsequent reactions. If the thiocyanate or cyanate is rapidly added to a strongly cooled solution of the lithiated sulfur compound, high yields of the adducts are obtained indicating that under these conditions the subsequent reactions are insignificant.

The methyne proton in the carbamoylation product obtained from $(CH_3S)_2CHLi$ and $(CH_3)_2NC(=O)Cl$, $(CH_3S)_2CHC(=O)N(CH_3)_2$, is very acidic. When a solution of the lithiated S,S-acetal in THF was added to a 200 mol% excess of dimethylcarbamoyl chloride at low temperatures, the yield of the carboxamide was only 40%. Introduction of a COOR group with ClCOOR has been reported by Seebach et al. [1], but a many-fold excess of chloroformate has to be used in order to obtain acceptable yields of the esters $(RS)_2CHCOOR$. With 200 mol% excess we obtained very poor results.

Procedure

A solution of 0.05 mol of lithiated bis(methylthio)methane in THF and hexane (Exp. 1) is cooled to $-80\,°C$, after which a mixture of 0.05 mol (3.7 g) of methyl isothiocyanate and 30 ml of THF is added over few seconds with vigorous stirring. The cooling bath is removed and the temperature allowed to rise to $-20\,°C$. Subsequently, cold 2 M hydrochloric acid (corresponding to ~ 0.05 mol) is added with vigorous stirring. After separation of the layers and extraction of the aqueous layer with Et_2O (four times), the organic solution is dried over $MgSO_4$. Concentration in vacuo gives the reasonably pure thioamide in an excellent yield. Crystallization from Et_2O (at $-25\,°C$) gives a m.p. 74.7–75.0 °C.

Addition at -70 to $-40\,°C$ (over 10 min) of a solution of 0.10 mol of $LiCH(SCH_3)_2$ (Exp. 1) to a mixture of 0.30 mol (32.3 g) of N,N-dimethylcarbamoyl chloride and 50 ml of THF gave, after the usual work-up and removal of the excess of $ClC(=O)NMe_2$ and unconverted $H_2C(SCH_3)_2$ by vacuum distillation, the expected carboxamide as a white solid. Crystallization from Et_2O (at $-25\,°C$) gave the pure product (m.p. 96.8–97.2 °C) in only $\sim 45\%$.

With $ClCOOC_2H_5$ a similar procedure gave a viscous liquid from which the desired ester could be isolated in $\sim 10\%$ yield by vacuum distillation.

Literature

1. Seebach D (1969) Synthesis 17

3.12 Peterson-Olefination Reactions with Lithiated Trimethylsilylbis(methylthio)methane. Preparation of Ketene Thioacetals

$$LiC(SiMe_3)(SCH_3)_2 + PhCH=O \xrightarrow[-10 \to +20\,°C]{THF\text{-}hexane} PhCH=C(SCH_3)_2 + Me_3SiOLi$$

$$LiC(SiMe_3)(SCH_3)_2 + C_2H_5CH=O \longrightarrow C_2H_5CH=C(SCH_3)_2 + Me_3SiOLi$$

$$LiC(SiMe_3)(SCH_3)_2 + cyclohexanone \longrightarrow (CH_2)_5C=C(SCH_3)_2 + Me_3SiOLi$$

Apparatus: p. 24, Fig. 1; 500 ml

Scale: 0.05 molar

Introduction

Seebach and co-workers successfully applied the Peterson olefination reaction (reviewed by Ager [1]) to prepare a number of ketene-S,S-acetals from carbonyl compounds and trimethylsilylated S,S-acetals [2]. The usual procedure involves addition of the carbonyl compound at low temperature to the lithiated Me_3Si—S,S-acetal, after which the temperature is allowed to rise over several hours to + 20 °C. In many cases these times presumably can be shortened considerably as we observed a strong heating effect upon addition of the carbonyl compounds to the lithiated S,S-acetal: the rise of the temperature was much greater than that usually observed in α-hydroxyalkylation reactions carried out with the same molar amounts of reagents and with the same concentration.

Procedure

A mixture of the silylated S,S-acetal (0.05 mol, 9.0 g, see Exp. 9) and 50 ml of THF is cooled to − 60 °C, after which a solution of 0.055 mol of BuLi in ~ 35 ml of hexane is added over a few s from a syringe. The cooling bath is removed and the temperature allowed to rise to 0 °C. After an additional 15 min the light-yellow solution is cooled to − 60 °C and 0.055 mol of the carbonyl compound is added in one portion with vigorous stirring. The cooling bath is removed immediately and the temperature rises to 0–10 °C within a few s. The colourless solution is warmed for 15 min in a bath at 35 °C, after which ice water (100 ml) is added with vigorous stirring. The layers are separated and the aqueous layer extracted three times with small portions of Et_2O or pentane. After drying the organic solution over magnesium sulfate, the solvent is removed under reduced pressure and the remaining liquid carefully distilled through a 30-cm Vigreux column. The following ketene-S,S-acetals were prepared with > 80% yields:

$PhCH\!=\!C(SCH_3)_2$, b.p. ~ 70 °C/0.5 mmHg;
$C_2H_5CH\!=\!C(SCH_3)_2$, b.p. 78 °C/12 mmHg, $n_D(20)$ 1.5197;
$(CH_2)_5C\!=\!C(SCH_3)_2$, b.p. 125 °C/12 mmHg, $n_D(20)$ 1.5659.

Literature

1. Ager DJ (1984) Synthesis 384
2. Gröbel BT, Seebach D (1977) Synthesis 390

3.13 Conjugate Addition of Lithiated S,S-Acetals and Corresponding S-Oxides to 2-Cyclohexen-1-one and Methylacrylate

$$R-C-(SCH_3)_2$$

$$LiC(R)(SCH_3)_2 + \quad [\text{cyclohexenone}] =O \quad \xrightarrow{\text{THF-hexane}} \quad [\text{cyclohexanone with } R-C-(SCH_3)_2] =O \text{ (after hydrolysis)}$$

R = Ph or SiMe₃

$$LiCH(SEt)(S\text{-}Et) + H_2C=CHCOOCH_3 \xrightarrow{\text{THF-hexane}} (EtS)(S\text{-}Et)CHCH_2CH_2COOCH_3$$

(with =O below LiCH group and =O below product) (after hydrolysis)

Apparatus: p. 24, Fig. 1; 500 ml

Scale: 0.05 molar

Introduction

The possibility of conjugate addition of metallated S,S-acetals and corresponding S-oxides to α,β-unsaturated carbonyl and nitro compounds has been investigated extensively [1]. Although in the reaction with C=O compounds 1,4-addition is the thermodynamically preferred process, in many cases addition to the C=O group occurs predominantly or exclusively. The outcome (C=O or 1,4-addition) of a reaction of a lithiated S,S-acetal or S-oxide is difficult to predict on a rational basis, but it seems that the tendency of the metallated sulfur compounds to enter into 1,4-addition becomes stronger when groups are present that stabilize negative charge. Thus $LiC(SiMe_3)(SCH_3)_2$ and some lithiated carboxylates $LiC(COOCH_3)(SR)_2$ give only 1,4-adducts with cyclopentenone and cyclohexenone. With these data on hand it is difficult to understand why $LiCH(SOC_2H_5)(SC_2H_5)$ and α,β-unsaturated ketones give the 1,2 (or C=O) adduct, and interaction between the *ethyl* compound $LiC(C_2H_5)(SOC_2H_5)(SC_2H_5)$ and these ketones results in the exclusive 1,4-addition [2]. Ostrowski and Kane [3] showed that at very low temperatures and in the presence of relatively small amounts of THF, 2-lithio-2-phenyl-1,3-dithiane and some cyclic enones coupled in a kinetically preferred 1,2-fashion. When the initial reaction mixture was kept at room temperature for 1 hour, only 1,4-adduct could be isolated in excellent yield after aqueous work-up. In these cases, the formation of the C=O adduct is a reversible reaction. Seebach et al. were unable to detect any kinetic formation of a 1,2-adduct at the very low temperature of − 100 °C in the reaction of 2-lithio-2-trimethylsilyl-1,3-dithiane with cyclohexenone [1].

The procedures described in this experiment are examples of 'clean' 1,4-additions. It should be stressed that they are only representative with respect to the experimental conditions.

Procedure

1. Generation of the lithium compounds

A solution of 0.055 mol of BuLi in ~ 35 ml of hexane is placed in the flask. THF (50 ml) is added with cooling below 0 °C. The mixture is brought at a temperature of − 40 °C, after which $PhCH(SCH_3)_2$ or $Me_3SiCH(SCH_3)_2$ (0.05 mol, for the preparation see Exps. 23 and 9) is added in one portion. The cooling bath is removed and the temperature allowed to rise to 0 °C. After an additional 5 or 20 min (respectively) at 0 °C the metallation is considered complete.

The S-oxide (0.05 mol, 7.6 g) and THF (50 ml) are placed in the flask and the mixture is cooled to − 60 °C. A solution of 0.055 mol of BuLi in ~ 35 ml of hexane is then added over a few s from a syringe. The lithiation occurs almost instantaneously.

2. Reaction with cyclohexenone and methyl acrylate

The solution of the lithium compound is cooled to − 60 °C and 0.055 mol (5.3 g) of cyclohexenone or freshly distilled methyl acrylate (4.7 g) is added in one portion, followed by immediate removal of the cooling bath. The temperature rises in the three cases to ca. − 25 °C within a few s, while in the case of $PhC(Li)(SCH_3)_2$ the characteristic red colour discharges immediately. The almost clear solutions are allowed to warm up to room temperature, after which stirring is continued for 30 min. The reaction mixtures obtained from the addition of $PhC(Li)(SCH_3)_2$ and $Me_3SiC(Li)(SCH_3)_2$ are worked up by addition of water, extraction of the aqueous layers with Et_2O, drying of the organic solutions over $MgSO_4$, and removal of the solvents in vacuo. The solid residues are crystallized from Et_2O to give the pure 1,4-adducts in greater than 75% yields.

The reaction mixture obtained from the lithiated S-oxide and methyl acrylate is worked up by removing the greater part of the THF and hexane under reduced pressure, adding water (100 ml) and extracting four times with chloroform. The extracts are dried over $MgSO_4$ and subsequently concentrated in vacuo, the last traces of chloroform being removed at < 1 mmHg pressure. The residue (yield almost quantitative) is reasonably pure (^1H NMR) 1,4-adduct.

Literature

1. Gröbel BT, Seebach D (1977) Synthesis 380; Bürstinghaus R, Seebach D (1977) Chem Ber 110:841
2. Herrmann JL, Richman JE, Schlessinger RH (1973) Tetrahedron Lett 3271
3. Ostrowski PC, Kane VV (1977) Tetrahedron Lett 3549

3.14 Lithiation of Dimethyl Sulfide and Methyl Phenyl Sulfide and Subsequent Reaction of the Lithium Compounds with Benzaldehyde and Trimethylchlorosilane

$$CH_3SCH_3 + BuLi \cdot TMEDA \xrightarrow[20 \to 40\,^\circ C]{hexane} LiCH_2SCH_3 \cdot TMEDA\downarrow + C_4H_{10}$$

$$LiCH_2SCH_3 + PhCH{=}O \xrightarrow[-60 \to -20\,^\circ C]{THF\text{-}hexane} PhCH(OH)CH_2SCH_3$$

(after hydrolysis)

$$CH_3SPh + BuLi \cdot TMEDA \xrightarrow[20 \to 40\,^\circ C]{hexane} LiCH_2SPh \cdot TMEDA\downarrow + C_4H_{10}$$

$$LiCH_2SPh + Me_3SiCl \xrightarrow{Et_2O\text{-}hexane} Me_3SiCH_2SPh + LiCl$$

Apparatus: p. 24, Fig. 1; 11

Scale: 0.10 molar

Introduction

There are only a few reports on successful metallations of simple saturated sulfides [1–3]. Lithiation of dimethyl sulfide has been achieved with BuLi.TMEDA in hexane [1], while methyl phenyl sulfide can be specifically converted into LiCH$_2$SPh using BuLi.TMEDA or BuLi.DABCO (1,4-diazabicyclo[2, 2, 2]octane). These reagents do not give good results in attempts to lithiate the homologues, RCH$_2$SPh.

An excess of CH$_3$SCH$_3$ is used to ensure the complete conversion of BuLi: reaction with PhCH$=$O would give rise to PhCH(OH)C$_4$H$_9$, which is difficult to separate from the desired compound. In the case of PhSCH$_3$, an excess of BuLi may cause subsequent ring-metallation [4], and similar difficulties may arise during the isolation of the silylation product.

Procedure

A mixture of 0.20 mol of dimethyl sulfide (12.4 g, 100 mol% excess) or 0.10 mol (12.4 g) methyl phenyl sulfide and 50 ml of hexane is placed in the flask. Dry TMEDA (0.12 mol, 13.9 g) is added, after which a solution of 0.10 mol of BuLi in ∼ 67 ml of hexane (in the case of CH$_3$SCH$_3$) or 0.10 mol of BuLi in ∼ 67 ml of hexane (in the case of CH$_3$SPh) is introduced over 5 min from a syringe. The temperature rises over a few min to 35 or 40 °C. The metallations are terminated by heating for 30 min in a water bath at 40 °C. After cooling to below − 10 °C (thick suspensions are formed) THF (50 ml) is added and the obtained solutions are immediately cooled to − 60 °C. Benzaldehyde (0.10 mol, 10.6 g) and freshly distilled trimethylchlorosilane (0.13 mol, 14.0 g), respectively, are added in one portion with vigorous stirring. The cooling bath is removed and, after an additional 10 min, 100 ml of ice water is added with vigorous stirring. The organic layers and two ethereal extracts are combined and dried over $MgSO_4$. The liquid remaining after removal of the volatile components in vacuo is distilled through a 20-cm Vigreux column to give $CH_3SCH_2CH(OH)Ph$, b.p. 100 °C/0.5 mmHg, $n_D(20)$ 1.5649, and $PhSCH_2SiMe_3$, b.p. 110 °C/12 mmHg, $n_D(20)$ 1.5392, in excellent yields.

Literature

1. Peterson DJ (1967) J Org Chem 32:1717
2. Corey EJ, Seebach D (1966) J Org Chem 31:4097
3. Dolak TM, Bryson TA (1977) Tetrahedron Lett 1961
4. Cabiddu S, Floris C, Melis S, Sotgiu F (1984) Phosphorus and Sulfur 19:61

3.15 Lithiation of (Trimethylsilylmethyl) Phenyl Sulfide and Subsequent Reaction with Acetone

$$PhSCH_2SiMe_3 + BuLi \xrightarrow[0\to30\,°C]{THF\text{-}hexane} PhSCH(Li)SiMe_3 + C_4H_{10}$$

$$PhS-CH(Li)SiMe_3 + (CH_3)_2C=O$$

$$\xrightarrow{THF\text{-}hexane} PhS-CH(SiMe_3)C(CH_3)_2OLi$$

$$\longrightarrow PhS-CH=C(CH_3)_2 + Me_3SiOLi$$

Apparatus: p. 24, Fig. 1; 500 ml

Scale: 0.05 molar

Introduction: see Exp. 12

Procedure

A solution of 0.055 mol of butyllithium in 36 ml of hexane and 40 ml of THF are added together with cooling below 0 °C. The mixture is cooled to − 10 °C, after which 0.05 mol (9.8 g) of (trimethylsilylmethyl) phenyl sulfide (see Exp. 14) is added. The cooling bath is removed. Above + 5 °C the reaction is markedly exothermic. After the evolution of heat has ceased, the yellow solution is warmed for an additional 10 min at 25 to 30 °C and subsequently cooled to − 60 °C. Dry acetone (0.06 mol, ∼ 3.5 g) is added in one portion after the cooling bath has been removed. The temperature of the solution rises within a few s to about 0 °C. After an additional 10 min 50 ml of ice water is added with vigorous stirring. After separation of the layers two extractions with small portions of Et_2O are carried out. The combined organic solutions (almost colourless) are dried over $MgSO_4$ and subsequently concentrated under reduced pressure, the last traces of volatile compounds being removed at 1 mmHg or lower pressure (bath temperature $\sim 20\,°C$). The remaining liquid is almost pure vinylic sulfide. The yield is almost quantitative.

3.16 Dilithiation of Methyl Phenyl Sulfide and Subsequent Trimethylsilylation

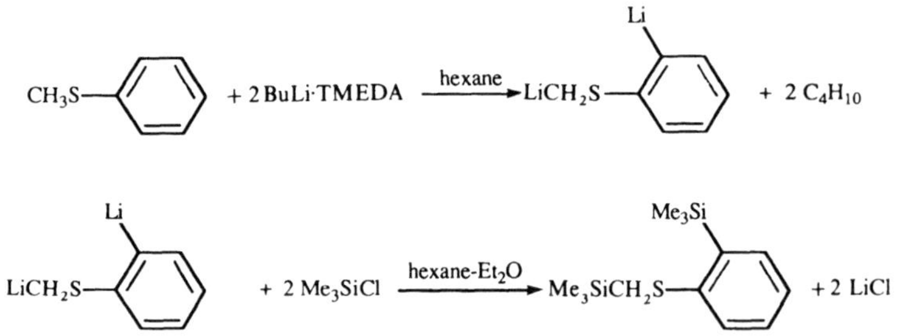

Apparatus: p. 24, Fig. 1; 500 ml

Scale: 0.05 molar ($PhSCH_3$)

Introduction

As shown in Exp. 14, methyl phenyl sulfide can be successfully converted into $LiCH_2SPh$ with one mol equivalent of the BuLi·TMEDA reagent. Very recently Cabiddu et al. [1] reported that the mono-lithium compound can be specifically lithiated at the *ortho*-carbon atom in the ring by prolonged treatment with a second equivalent or butyllithium. A useful synthetic application is the preparation of benzo[b]thiophenes by reaction of the dilithio compounds with acyl chlorides [2].

In the procedure described below, the formation of the dilithiated sulfide is demonstrated by the reaction with trimethylchlorosilane. The reaction time can be shortened considerably by carrying out the dimetallation at elevated temperatures and using hexane as the only solvent.

Procedure

A mixture of 0.05 mol (6.2 g) of methyl phenyl sulfide and 0.11 mol of TMEDA is placed in the flask. A solution of 0.06 mol (note 1) of butyllithium in 36 ml of hexane is added over a few s from a syringe. The temperature of the solution rises within a few min to ca. 40 °C and is kept at this level for an additional 15 min. An additional amount of 0.05 mol of BuLi in 32 ml of hexane is then added and the mixture is heated at 55 °C for 45 min. A rather thick suspension is formed. After cooling to

$-30\,°C$ 0.12 mol (13.0 g) of freshly distilled trimethylchlorosilane is added with virorous stirring. The cooling bath is removed and the temperature allowed to rise to $+10\,°C$. Cold (0 °C) dilute hydrochloric acid is added with stirring in order to reach pH < 3. The organic layer and one ethereal extract are washed with water and dried over $MgSO_4$. The liquid remaining after concentration of the organic solution in vacuo is distilled through a 5-cm Vigreux column to give the disilyl derivative, b.p. 145 °C/12 mmHg, $n_D(20)$ 1.5253, in $\sim 90\%$ yield.

Notes

1. If necessary in connection with subsequent functionalization reactions, the excess of BuLi can be made inactive by adding a sufficient amount of THF (~ 50 ml) which reacts with BuLi·TMEDA at room temperature to give ethene and $H_2C{=}CHOLi$.

Literature

1. Cabiddu S, Floris C, Melis S, Sotgiu F (1984) Phosphorus and Sulfur 19:61
2. Cabiddu S, Cancellu D, Floris C, Gelli G, Melis S (1988) Synthesis 888

3.17 Reaction of Bis(methylthio)methane with Potassium Amide in Liquid Ammonia and Subsequent Reaction with Oxirane

$$H_2C(SCH_3)_2 + KNH_2 \xrightarrow[-33\,°C]{\text{liq. NH}_3} KCH(SCH_3)_2 + NH_3$$

$$KCH(SCH_3)_2 + \text{oxirane} \xrightarrow[-40\,°C]{} KOCH_2CH_2CH(SCH_3)_2$$

$$\xrightarrow{H_2O} HOCH_2CH_2CH(SCH_3)_2$$

Apparatus: 1-l round-bottomed, three-necked flask (vertical necks!) equipped with a mechanical stirrer, a dropping funnel and a thermometer combined with a vent (the apparatus of p. 24, Fig. 1; can be used).

Scale: 0.30 molar

Introduction

The conditions described in this experiment are typical for hydroxyalkylation of anionic species with oxirane in liquid ammonia. The reaction with 'sp³'-anions in this solvent is very fast, except in the case of enolates. Yield of the β-hydroxyalkyl derivatives are good, provided the anionic species is present in a sufficiently high equilibrium concentration. This is probably not the case when *lithium* amide is used for the generation of the anion $^-CH(SCH_3)_2$ (the pK values of the S,S-acetal and ammonia are presumably very close to each other). With sodamide the 'ionization' of $H_2C(SCH_3)_2$ may be incomplete, but if the epoxide is added at a sufficiently low rate, it may preferentially react with the soluble $NaCH(SCH_3)_2$, thus giving the (fast) equilibrium the opportunity to shift to the right side. Other epoxides, e.g., epoxypropane and epoxystyrene will probably also give good results in their reactions with the S,S-acetal anion in liquid ammonia.

Procedure

In the flask is prepared a solution of 0.35 mol of potassium amide in 300 ml of liquid ammonia (see Vol. 1, Chap. I). Bis(methylthio)methane (0.30 mol, 32.4 g, see Exp. 21) is added over 5 min, then the greenish solution is cooled to $\sim -40\,°C$ (cooling bath with CO_2 and acetone) while a stream of N_2 (~ 500 ml/min) is passed through the flask. A mixture of 0.50 mol (22 g) of oxirane and 50 ml of Et_2O (precooled to $-30\,°C$) is added over 10 min with stirring at a moderate rate. Efficient cooling is necessary to keep the temperature between -35 and $-45\,°C$ (below the b.p. of ammonia). After an additional 10 min, the cooling bath, the dropping funnel, and the thermometer are removed. The flask is placed in a water bath at 35 °C. When the greater part of the ammonia has evaporated, dichloromethane (150 ml, note 1) and ice water (250 ml) are added in succession. The layers are separated (note 2) and the aqueous phase is extracted three times with dichloromethane. The combined organic solutions are dried over $MgSO_4$ and subsequently concentrated in vacuo. Distillation of the remaining liquid through a short column gives the alcohol in an excellent yield (see Exp. 4).

Notes

1. Et_2O should not be used, since the very small pieces of unconverted potassium, which are sometimes, present, may give rise to fire hazards.
2. If too much ferric nitrate is used for the preparation of potassium amide, the separation of the layers may be difficult due to the presence of a ferric hydroxide gel.

3.18 Reaction of Dimethylsulfoxide with Sodamide in Liquid Ammonia and Subsequent Alkylation with Bromohexane

$$CH_3SOCH_3 + NaNH_2 \xrightarrow[-33\,°C]{\text{liq. } NH_3} NaCH_2SOCH_3$$

$$NaCH_2SOCH_3 + C_6H_{13}Br \longrightarrow C_6H_{13}CH_2SOCH_3 + NaBr\downarrow$$

Apparatus: 1-l three-necked, round-bottomed flask, equipped with a dropping funnel, a mechanical stirrer and a vent.

Scale: 0.3 molar

Introduction

Addition of butyllithium or LDA to a solution of DMSO in THF gives a suspension of 'dimsyllithium'. Coupling with aldehydes and ketones gives the expected products in good yields. The reaction of this suspension with primary alkyl bromides proceeds less satisfactorily, since a mixture of comparable amounts of mono- and dialkylated (RCH_2SOCH_2R) product and DMSO is obtained. The best procedure for homologating DMSO consists of metallation with sodamide in liquid ammonia and subsequent addition of the alkyl halide. Under these conditions only minor quantities of dialkyl derivatives are formed. Other sulfoxides, of which the lithium compounds react sluggishly with alkyl halides in THF, e.g., $H_2C(SC_2H_5)(SOC_2H_5)$ and $H_2C{=}CHCH_2SOR$, and also sulfones, may be alkylated by a similar procedure.

Procedure

A suspension of 0.35 mol of sodamide in 500 ml of liquid ammonia is prepared as described in Vol. 1, Chap. I. To this suspension is added over 10 min a mixture of 0.30 mol (23.4 g) of dry DMSO and 50 ml of Et_2O. A greyish (colloidal iron) solution is formed. Subsequently, 0.30 mol (49.5 g) of bromohexane is added over 15 min to the vigorously stirred solution. After an additional 1 h, the ammonia is removed by placing the flask in a water bath at 40 °C. To the remaining salt mass is added 200 ml of water, after which seven extractions with chloroform ($1 \times 75 + 6 \times 25$ ml) are carried out. The unwashed extracts are dried over $MgSO_4$ and subsequently concentrated under reduced pressure. Careful distillation of the remaining liquid through a 30-cm Vigreux column gives the desired sulfoxide, b.p. $\sim 150\,°C/15\,mmHg$, in $\sim 70\%$ yield. The small residue consists mainly of $C_6H_{13}CH_2SOCH_2C_6H_{13}$.

3.19 Mono-Deuteration of Bis(ethylthio)methane

$$H_2C(SC_2H_5)_2 + BuLi \xrightarrow[-30 \to +10\,°C]{THF\text{-}hexane} LiCH(SC_2H_5)_2 + C_4H_{10}$$

$$LiCH(SC_2H_5)_2 + D_2O \longrightarrow DCH(SC_2H_5)_2 + LiOD\downarrow$$

Apparatus: p. 24, Fig. 1; 500 ml

Scale: 0.05 molar

Introduction

Deuterated compounds may be needed for mechanistic or physical studies and a very high purity (i.e., preferably at least 97% of the product should consist of the compound having the D-atom on the desired place) is often necessary. A general method consists in treating the alkali-metal derivative with deuterium oxide or deuteriomethanol or -ethanol. The presence of undeuterated compound may be due to incomplete metallation, therefore this reaction is carried out with an excess of the basic reagent. This involves, however, some risk of introduction of more than one D-atom, especially when the quench operation is not conducted in the proper way:

$$LiCH(SC_2H_5)_2 + D_2O \longrightarrow DCH(SC_2H_5)_2 + LiOD$$

$$DCH(SC_2H_5)_2 + BuLi \text{ (excess)} \longrightarrow DC(Li)(SC_2H_5)_2 \xrightarrow{D_2O} D_2C(SC_2H_5)_2$$

If the deuterating agent is slowly added to the solution of the lithium compound any remaining BuLi (from the excess) may remetallate $DCH(SC_2H_5)_2$, after which a second D-atom may be introduced. The chance on this complication can be minimized by adding a large excess of D_2O in one portion. Vigorous agitation during this addition may result in a quick and homogeneous distribution of the deuterating agent. A 100% excess is in many cases insufficient, since the alkali deuteroxide can absorb appreciable amounts of D_2O, which then are not available for the quench reaction. Too strong cooling during the addition of D_2O may lead to crystallization of part of this reagent. All possible complications can be avoided, however, by introducing the solution of the metallated compound (preferably with a syringe) into a vigorously agitated and cooled mixture (solution) of THF and a large (at least 500 mol%) excess of D_2O. Crystallization of the D_2O can be prevented by keeping the temperature around $0\,°C$.

Procedure (note 1)

To a mixture of 50 ml of THF and 0.05 mol (6.8 g) of bis(ethylthio)methane (see Exp. 21) is added (from a syringe) over a few min a solution of 0.06 mol of butyllithium in 36 ml of hexane. During this addition the temperature of the mixture is kept below $-20\,°C$. The cooling bath is removed and the temperature allowed to rise to $+5\,°C$. The solution is kept at this temperature for an additional 45 min, then it is added via a syringe to a vigorously agitated solution of 5 g (large excess) of D_2O in 50 ml of THF, kept between 0 and $-5\,°C$. This additional takes about 5 min. The white suspension or slurry is stirred for an additions 10 min, then water (50 ml) is added. The organic layer is dried over $MgSO_4$ and subsequently concentrated in vacuo. Distillation gives the pure (^1H NMR, GC-MS) mono-deuterated S,S-acetal, b.p. $68\,°C/12$ mmHg, $n_D(20)$ 1.5118.

Notes

1. A good procedure for the preparation of 2-phenyl-1,3-dithiane-2-d is described by Seebach D, Erickson BW, Singh G (1966) J Org Chem 31:4303

3.20 Lithiation of Methyl Phenyl Sulfoxide with LDA and Subsequent Alkylation with Butyl Bromide

$$PhSOCH_3 + LDA \xrightarrow[-20 \to +10°C]{THF\text{-}hexane} PhSOCH_2Li + HDA$$

$$PhSOCH_2Li + C_4H_9Br \longrightarrow PhSOCH_2C_4H_9 + LiBr$$

Apparatus: p. 24, Fig. 1; 500 ml

Scale: 0.05 molar

Introduction

Butyllithium is an unsuitable reagent for the lithiation of alkyl aryl sulfoxides, since it cleaves the Ph—S bond [1]. Specific metallation in the alkyl group can be readily achieved with lithium dialkylamides [2] in organic solvents and with alkali amides in liquid ammonia. Both types of bases are sufficiently strong to bring about complete metallation. The metallation with the two base-solvent systems offers complementary possibilities for coupling with the various electrophilic reagents. Although liquid ammonia is the best medium for reactions with alkylating agents, the procedure in this experiment shows that primary alkyl bromides react satisfactorily with lithiomethyl phenyl sulfoxide in a THF-hexane mixture.

Procedure

Methyl phenyl sulfoxide (0.05 mol, 7.0 g, see Exp. 22) is added in one portion to a solution of 0.06 mol of LDA in 37 ml of hexane and 50 ml of THF, cooled at −20 °C. The cooling bath is removed and the temperature allowed to rise. After an additional 30 min (at +10 °C) 0.2 mol (28 g) of butyl bromide is added in one portion to the solution (cooled to 0 °C). The temperature rises gradually to between 30 and 40°C. After an additional half hour (at 40 °C) the greater part of the THF and hexane is removed (rotary evaporator) under reduced pressure. After addition of 100 ml of water to the remaining viscous liquid, four extractions with 30-ml portions of chloroform are carried out. The unwashed organic solutions are dried over $MgSO_4$ and subsequently concentrated under reduced pressure. The last traces of volatile components are removed in a vacuum of < 1 mmHg. The remaining liquid ($n_D(20)$ 1.5363) appears to be pure pentyl phenyl sulfoxide. The yield is almost quantitative.

Literature

1. Lockard JP, Schroeck CW, Johnson CR (1973) Synthesis 485
2. Tsuchihashi G, Iriuchijima S, Maniwa K (1973) Tetrahedron Lett 3389

3.21 Preparation of Formaldehyde-S,S-Acetals

$$CH_3SSCH_3 + 2Na \xrightarrow{\text{liq. } NH_3} 2CH_3SNa$$

$$2CH_3SNa + CH_2Cl_2 \xrightarrow{\text{liq. } NH_3} H_2C(SCH_3)_2 + 2NaCl \downarrow$$

$$RSH + KOH \xrightarrow{C_2H_5OH} RSK + H_2O$$

$$2RSK + CH_2Cl_2 \xrightarrow{C_2H_5OH} H_2C(SR)_2 + 2KCl \downarrow \quad (R=C_2H_5 \text{ or } Ph)$$

Apparatus

For the reaction in liquid ammonia: 2-l three-necked, round-bottomed flask, equipped with a dropping funnel, an effcient mechanical stirrer and a vent. For the reactions in ethanol: 1-l three-necked round-bottomed flask, equipped with a dropping funnel, a mechanical stirrer and a reflux condenser.

Scale: 1.0 molar (CH_3SSCH_3 or RSH)

Procedure

Bis(methylthio)methane

Anhydrous liquid ammonia (1.5 l, Note 1) is placed in the flask. Sodium (2.0 mol, ~48 g, cut into pieces of about 1 g and freed from paraffin oil or high-boiling petroleum ether by rinsing with Et_2O) is then added. The dissolution takes about 10 min. Dimethyl disulfide (1.0 mol, 94 g) is added over 30 min with efficient stirring. The reaction is very vigorous. Slightly less or more than the 94 g may be needed to cause complete decolorization. Dichloromethane (1.0 mol, 85 g) is subsequently added dropwise over 30 min. Thirty min after this addition an additional amount of 30 g of dichloromethane is added (partly as compensation for losses due to evaporation along with the ammonia). If the volume of the mixture has decreased to less than 0.5 l, about 200 ml of ammonia is added. Stirring is continued for an additional 2 h. The ammonia is allowed to evaporate overnight. After addition of 500 ml of water and dissolution of the salt, five extractions with small portions of Et_2O are carried out. The extracts are dried over $MgSO_4$, after which the greater part of the Et_2O is distilled off under atmospheric pressure through a 30-cm Vigreux column. Subsequent vacuum distillation gives the S,S-acetal, b.p. 40 °C/12 mmHg, $n_D(20)$ 1.5338, in ~90% yield (Note 2).

Bis(ethylthio)methane and bis(phenylthio)methane

Ethanol (500 ml, 96%) and potassium hydroxide pellets (technical grade, 80 g) are placed in the flask. After dissolution of the pellets, the solution is cooled in a bath with ice and water and freshly distilled (Note 3) ethanethiol (1.0 mol, 62 g) or thiophenol. (1.0 mol, 110 g) is added over 10 min without external cooling. Subsequently, dichloromethane (0.5 mol, 42.5 g) is added over a few min, after which the mixture is heated under reflux for 1.5 h (after 1 h an additional amount of 20 g of dichloromethane is added). The mixture is cooled to below 30 °C, then poured into 2 l of water, after which extraction with dichloromethane is carried out. The organic solutions are washed with water, dried over $MgSO_4$ and subsequently concentrated under reduced pressure. Bis(ethylthio)methane is isolated by distillation through a 30-cm Vigreux column: b.p. 65 °C/12 mmHg, $n_D(20)$ 1.5124. Bis(phenylthio)methane is purified by crystallization from 96% ethanol: m.p. 40 °C. Both S,S-acetals are obtained in high yields (Note 4).

Notes

1. Full details about working with liquid ammonia are given in Brandsma L, Verkruijsse HD (1987) Preparative Polar Organometallic Chemistry, Vol. 1, Chap. I, Springer-Verlag and in Brandsma L (1988) Preparative Acetylenic Chemistry, Revised edition, Elsevier.
2. Bis(methylthio)methane may also be prepared by heating a solution of CH_3SK in methanol with dichloromethane (compare the procedure of $H_2C(SC_2H_5)_2$). The solution of CH_3SK can be prepared by adding a cold mixture of CH_3SH and CH_3OH to a solution of KOH in methanol.
3. The undistilled products may contain some disulfide, RSSR.
4. Attempts to prepare 1,3-dithiane from $HS(CH_2)_3SH$, KOH and dichloromethane gave only amorphous polymeric products.

3.22 Preparation of Ethylthiomethyl Ethyl Sulfoxide and Methyl Phenyl Sulfoxide

$$PhSCH_3 + H_2O_2 \xrightarrow[10 \to 20\,°C]{CH_3COOH} PhS(=O)CH_3 + H_2O$$

$$H_2C(SC_2H_5)_2 + H_2O_2 \xrightarrow[10 \to 25\,°C]{CH_3COOH} H_5C_2SCH_2S(=O)C_2H_5 + H_2O$$

Apparatus: 1-l round-bottomed, three-necked flask, equipped with a dropping funnel, a mechanical stirrer and a thermometer-outlet combination.

Scale: 0.5 molar

Procedure

Acetic acid (99 or 100%, 150 ml) and the sulfur compound (0.50 mol) are placed in the flask. The mixture is cooled to 10 °C, after which a 35% (by weight) aqueous solution of hydrogen peroxide (50 g, corresponding to 0.50 mol) is added dropwise over 45 min, while keeping the temperature between 15 and 20 °C (a bath with dry ice and acetone is recommended). After the addition, the cooling bath is removed and the temperature may slowly rise to 25 or 30 °C. The almost colourless solution is stirred for an additional 2 h at 25 °C, then the greater part of the acetic acid is removed under reduced pressure using a rotary evaporator. To the remaining liquid is added 100 ml of water, after which ten extractions with small (1×50 and 9×25) portions of chloroform are carried out. The combined, unwashed organic solutions are dried over $MgSO_4$ and subsequently concentrated under reduced pressure. Distillation through a short column gives $PhSOCH_3$, b.p. $\sim 100\,°C/0.5\,mmHg$, $n_D(20)$ 1.5362 and $H_2C(SC_2H_5)(SOC_2H_5)$, b.p. $\sim 90\,°C/0.3\,mmHg$, $n_D(20)$ 1.5280, in greater than 80% yields.

3.23 Preparation of Benzaldehyde Dimethylthioacetal

$$PhCH=O + 2CH_3SH \xrightarrow[10\to30\,°C]{ZnCl_2} PhCH(SCH_3)_2 + H_2O$$

Apparatus: 1-l three-necked, round-bottomed flask, equipped with a gas-inlet tube, a mechanical stirrer, and a thermometer-outlet combination. The outlet is connected with a trap cooled at $-78\,°C$.

Scale: 1.0 molar (benzaldehyde)

Procedure

Benzaldehyde (1.0 mol, 106 g, freshly distilled) and anhydrous, powdered zinc chloride (10.0 g) are placed in the flask. A trap in which 2.2 mol (106 g) methanethiol has been condensed is connected with the inlet tube. The trap is placed in a water bath at 15 to 20 °C. During the introduction of the thiol the mixture is vigorously agitated and the temperature kept (bath with dry ice and acetone) between 20 and 25 °C. When, after ~ 1.5 h, all methanethiol has evaporated (if the other trap contains some thiol, the traps are interchanged first), the mixture is cooled to 15 °C and an additional amount of 10.0 g of zinc chloride is added to the turbid mixture. Stirring is continued for 1.5 h at 30 °C, then the mixture is cooled to 0 °C and 300 ml of an aqueous solution of 40 g of potassium hydroxide is added with vigorous stirring. The layers are separated (addition of 100 ml of Et_2O may facilitate the separation), the organic layer is washed with dilute aqueous potassium hydroxide and subsequently dried over potassium carbonate. Distillation through a 20-cm Vigreux column gives, after a first fraction of benzaldehyde, the S,S-acetal, b.p. 145 °C/12 mmHg, $n_D(20)$ 1.5191, in a good to excellent yield.

4 Selected Procedures from Literature

Seebach D, Beck AK (1971) Org Synth 51:76

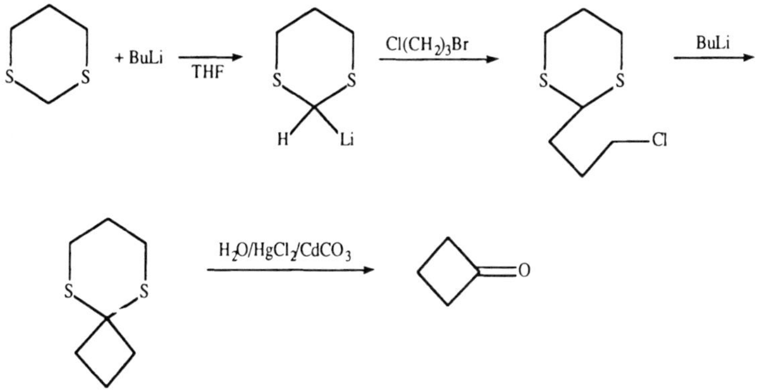

Seebach D, Beck AK (1971) Org Synth 51:39

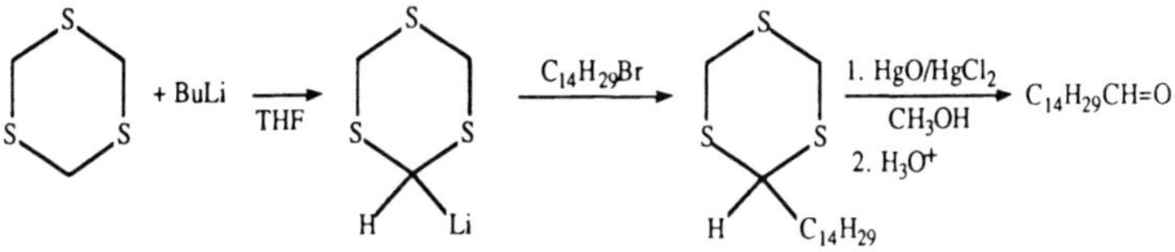

Stütz P, Stadler PA (1977) Org Synth 56:8

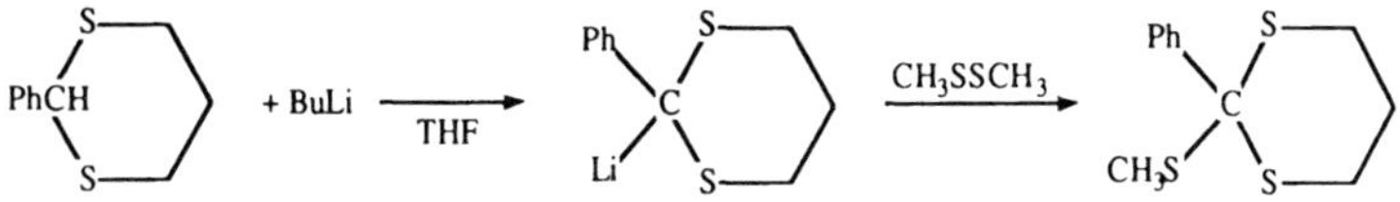

Photis JM, Paquette LA (1977) Org Synth 57:53

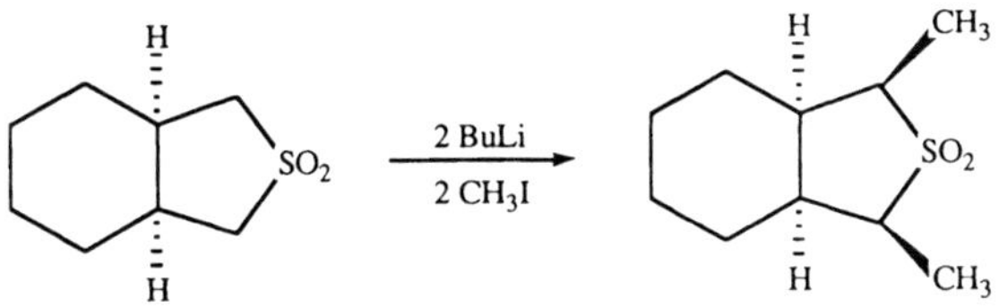

Cohen T, Ruffner RJ, Shull DW, Fogel ER, Falck JR (1979) Org Synth 59:203

$$H_2C(SPh)_2 \xrightarrow[THF]{BuLi} LiCH(SPh)_2 \xrightarrow{H_2C=CHCH=O}$$

$$H_2C=CHCH(OH)CH(SPh)_2$$

Martin SF (1979) Synthesis 640

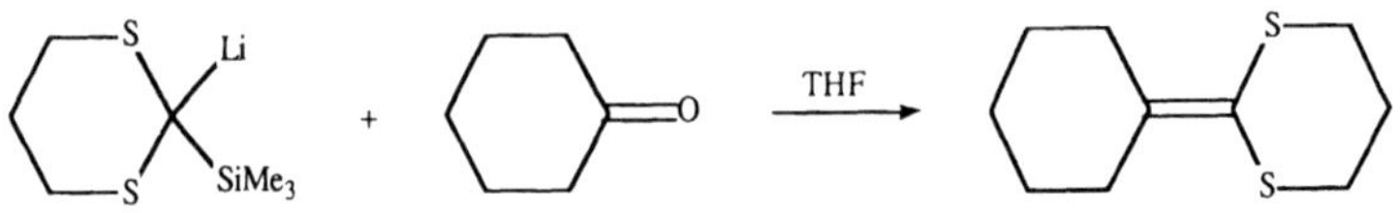

Seebach D (1969) Synthesis 32

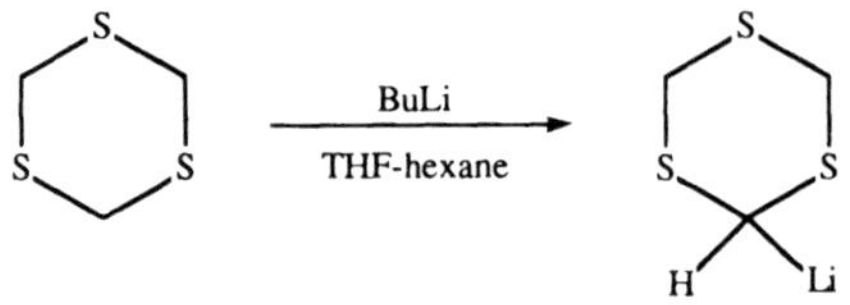

Gröbel BT, Seebach D (1977) Synthesis 380

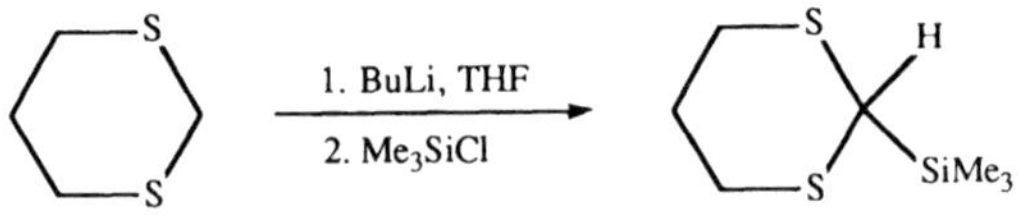

Chapter IV
α-Metallation of Derivatives
of Toluene Containing Heterosubstituents

1 Scope of this Chapter

A number of heterosubstituted toluenes have been shown to undergo exclusive metallation at the methyl group instead of *ortho*-metallation upon treatment with a strong base like butyllithium. This reaction has found some useful applications in the synthesis of condensed heterocyclic compounds [1]. The derivatives 1 and 2 having Me_2N and Me_2NCH_2 groups [2, 3] in the *ortho*-position of methyl presumably form a coordination complex with the base, after which intramolecular CH_3-metallation occurs. A similar mechanism may explain the metallation at the methyl group in 3 by (two equivalents of) butyllithium [4] where the second equivalent of base can form a mixed aggregate with the lithium alkoxide primarily formed, and at the methyl group in 4 [5] (initial formation of $N=C-OLi$ which can function as a coordinating group).

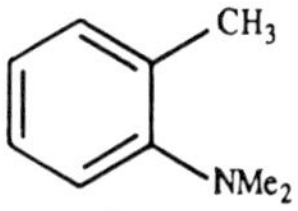

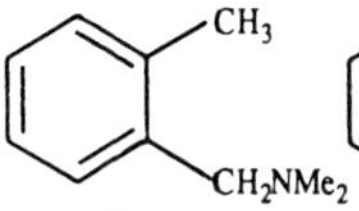

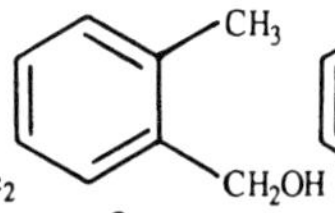

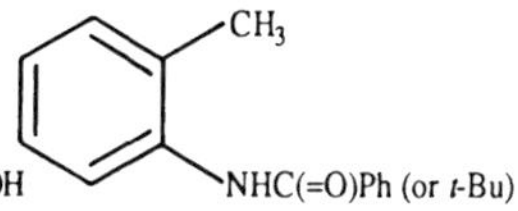

Although the strongly electron-withdrawing properties of the substituents in compounds 5–7 suggest a base like LDA to be sufficiently strong to effect complete CH_3-metallation, butyllithium seems in fact to be required [6–8].

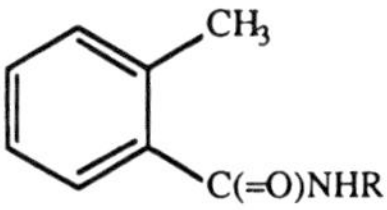

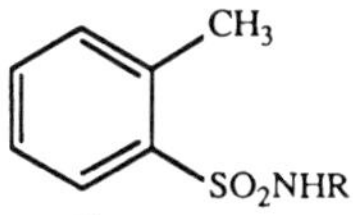

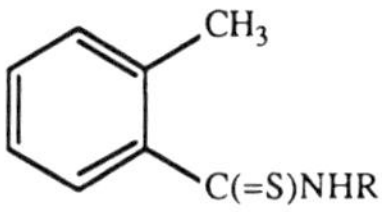

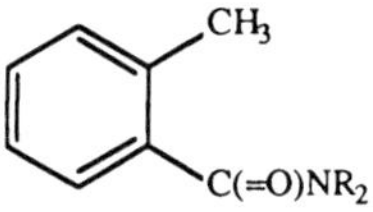

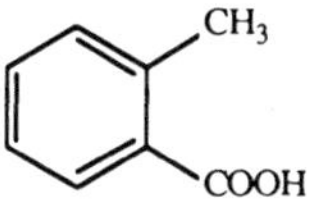

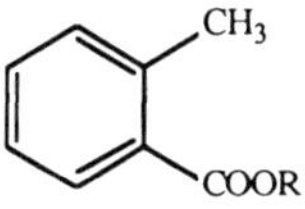

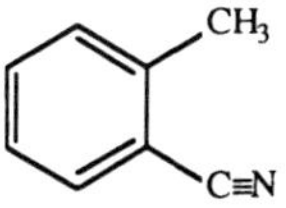

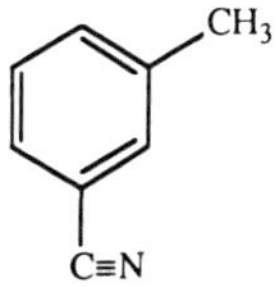

Lithium diisopropylamide has been successfully used for the lateral metallation of compounds 8–12 [9–12]. Even with the less strongly basic alkali amides in liquid ammonia the anions of *o* and *p*-tolunitrile can be generated in high equilibrium concentration as suggested by the good yields of the products obtained after addition of alkyl halides [13]. *Meta*-tolunitrile gave poor results in reactions with alkali amide and LDA. The thermodynamic acidity of 13 is probably much lower than that of 11 and 12 (pK difference of the order of 2 or 3), because conjugative stabilization in the anion is not optimal. As a consequence, the metallation is probably not complete and other reactions such as addition of the base or the nitrile anion across the C≡N function or one-electron transfer (in the case of LDA) may occur to a considerable extent. In the dimethylbenzoic acid 14 the methyl group in the *para*-position of the COOH-group is preferentially metallated [10].

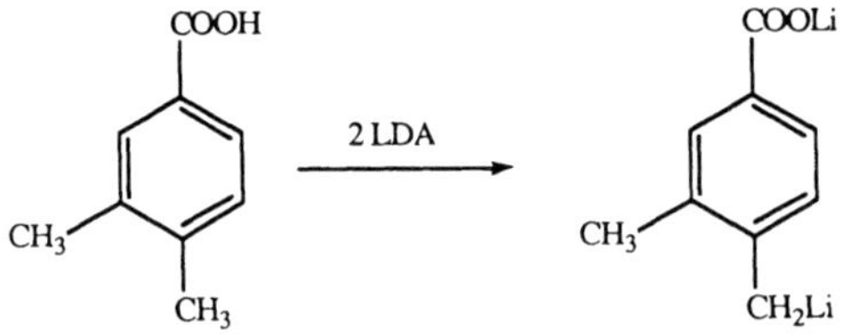

14

A group with strong inductively withdrawing properties can cause a considerable increase of the acidity of a methyl proton if the CH_3-group and this substituent are in suitable (*ortho-* or *para-*) relative positions. Compounds 15, for example, can be metallated with alkali amides in liquid ammonia or with LDA in organic solvents [13]:

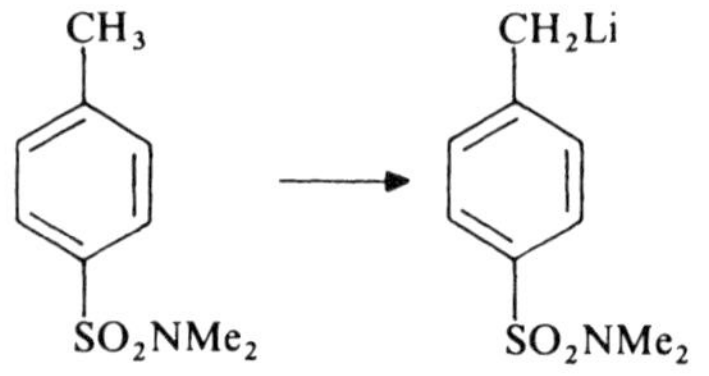

15

In analogy with the regiospecific metallation at the methyl group in 1 one would expect a similar result for the metallation of methyl *o*-cresyl ether 16. Treatment of 16 with the same bases (BuLi or BuLi·TMEDA) as used for the metallation of 1 initially gave mainly (> 90%) *ortho*-lithiated product [13]. Prolonged refluxing gives rise to a considerable decrease of the ratio of ring-metallated and CH_3-metallated derivatives [14]. Using a 1:1:1 molar mixture of BuLi, *t*-BuOK, and TMEDA in hexane, complete metallation could be achieved at −20°C within 1–2 hours, the predominant product now being the CH_3-metallated derivative [13].

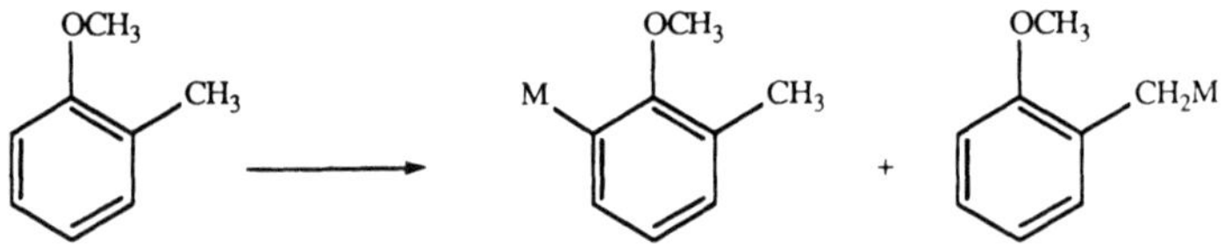

16

A recent report [15] deals with the dimetallation of *o*-, *m*-, and *p*-cresol (**17**) with a 1:1 molar mixture of BuLi and *t*-BuOK in inert solvents

Ortho- and *meta*-cresol are reported to give good yields but the dimetallation of *para*-cresol proceeded much less easily and gave lower yields of derivatization products. We obtained good results in the dimetallations of *o*- and *m*-cresol with 1:1:1 mixtures of BuLi, *t*-BuOK, and TMEDA at $-20\,°C$ [15].

Literature

1. Narasimhan NS, Mali RS (1987) in: Tropics in Current Chemistry 138:63, Springer-Verlag
2. Ludt RE, Crowther GP, Hauser CR (1970) J Org Chem 35:1288
3. Vaulx RL, Jones FN, Hauser CR (1964) J Org Chem 29:1387
4. Braun M, Ringer E (1983) Tetrahedron Lett 1233
5. Fuhrer W, Gschwend HW (1979) J Org Chem 44:1133; Houlihan WJ, Parrino VA, Uike Y (1981) J Org Chem 46:4511
6. Vaulx RL, Puterbaugh WH, Hauser CR (1964) J Org Chem 29:3514
7. Watanabe H, Hauser CR (1968) J Org Chem 33:4278
8. Fitt JJ, Gschwend HW (1976) J Org Chem 41:4029
9. Watanabe M, Sahara M, Furukawa S, Billedeau RJ, Snieckus V (1982) Tetrahedron Lett 1647; see also Watanabe M, Sahara M, Kubo M, Furukawa S, Billedeau RJ (1984) J Org Chem 49:742
10. Creger PL (1970) J Am Chem Soc 92:1396
11. Kraus GA (1981) J Org Chem 46:201
12. Kaiser EM, Petty JD (1976) J Organometal Chem 107:219
13. Heus-Kloos YA, Andringa H, Tip L, Brandsma L (unpublished)
14. Harmon TE, Shirley DA (1974) J Org Chem 39:3164
15. Bates RB, Siahaan TJ (1986) J Org Chem 51:1432

2 Experiments

All temperatures are internal, unless indicated otherwise.

For general instructions concerning handling organoalkali reagents, drying solvents, etc., see Vol. 1, Chap. I.

All reactions are carried out in an atmosphere of inert gas, except those in boiling ammonia.

2.1 Metallation of *N,N*-Dimethyl-*ortho*-Toluidine

Functionalization with trimethylchlorosilane.

Apparatus: p. 24, Fig. 1; 500 ml

Scale: 0.10 molar

Introduction

Hauser et al. (see Sect. 1, Ref. [2]) described the metallation of N,N-dimethyl-*ortho*-toluidine and the *para*-isomer with BuLi.TMEDA in hexane. Both isomers underwent regiospecific metallation, the *ortho*-isomer in the CH$_3$-group, the *para*-isomer in the ring, next to the NMe$_2$ group. The regiospecificity of the metallations may be explained by assuming the formation of initial coordination complexes with BuLi in which the observed lithiations take place. Relatively smooth CH$_3$- and ring-metallations, respectively, can be achieved only when TMEDA is present. Manzer [1] showed that CH$_3$-metallation in the *ortho*-isomer is possible in the absence of TMEDA, but at 20 °C in a mixture of hexane and Et$_2$O the time for completion is about two days. We [2] found that even at reflux temperature the metallation proceeds sluggishly. If an equivalent amount of TMEDA is added, however, the CH$_3$-metallation in the *ortho*-toluidine is complete within ca. 30 minutes at ~ 50 °C. It should be noted that this lateral lithiation is considerably faster than the *ortho*-lithiation of N,N-dimethylaniline (compare Vol. 1, Chap. VI). Potassiation of the methyl group in the *ortho*-toluidine can be achieved within one hour at about − 20 °C with a 1:1:1 molar mixture of BuLi, *t*-BuOK, and TMEDA in hexane. The high efficiency and regiospecificity appears after quenching with trimethyl-chlorosilane. The reaction conditions are similar to those for the potassiation of methylnaphthalene (Chap. II, Exp. 2).

Procedure

A solution of butyllithium (0.115 mol in ~ 75 ml of hexane) is placed in the flask. TMEDA (0.12 mol, 13.9 g) is added in one portion at room temperature, after 2 min followed by 0.1 mol (13.5 g) of N,N-dimethyl-*para*-toluidine (note 1). The reaction has a moderate heating effect as may be shown by placing the flask in cotton wool. The thermometer-outlet combination is replaced by a reflux condenser and the reaction mixture (orange suspension) heated for 30 min under gentle flux. The suspension is then cooled (the flow of inert gas is temporarily increased) to − 50 °C and 60 ml of Et$_2$O is added, followed by 0.13 mol (14.1 g) of freshly distilled trimethylchlorosilane. The cooling bath is removed. A light-yellow suspension is formed. After an additional 10 min, 100 ml of water is added with vigorous stirring. After separation of the layers, two extractions with Et$_2$O are carried out. The combined organic solutions are dried over K$_2$CO$_3$ and subsequently concentrated under reduced pressure. The silylation product, b.p. 100 °C/12 mmHg, is obtained in > 90% yield after distillation through a 30-cm Widmer column.

Notes

1. Preparation of *N,N*-dimethyl-*para*-toluidine.
   Dimethyl sulfate (140 ml) is added in one portion to a vigorously stirred mixture of 1.45 mol (155.2 g) of p-toluidine and 150 ml of water. The mixture is stirred over 12 h at room temperature, then the mixture is made slightly (pH ~ 8) alkaline by controlled addition of a 50% aqueous solution of potassium hydroxide. After cooling to room temperature, a second portion of 140 ml of dimethyl sulfate is added and the procedure is repeated. A third portion of dimethyl sulfate is added and after 12 h of stirring the reaction mixture is made strongly alkaline. The product is isolated by extraction with Et_2O, drying the organic solution over potassium carbonate, removal of the solvent under reduced pressure, and distillation of the remaining liquid through a 20-cm Widmer column. The yield of *N,N*-dimethyl-*para*-toluidine, b.p. ~ 45 °C/0.5 mmHg, is about 80%.

Literature

1. Manzer LE (1978) J Am Chem Soc 100: 8068
2. Heus-Kloos YA, Brandsma L (unpublished)

2.2 Synthesis of *ortho*-Pentylphenol *via* Potassiation of O-Protected *ortho*-Cresol

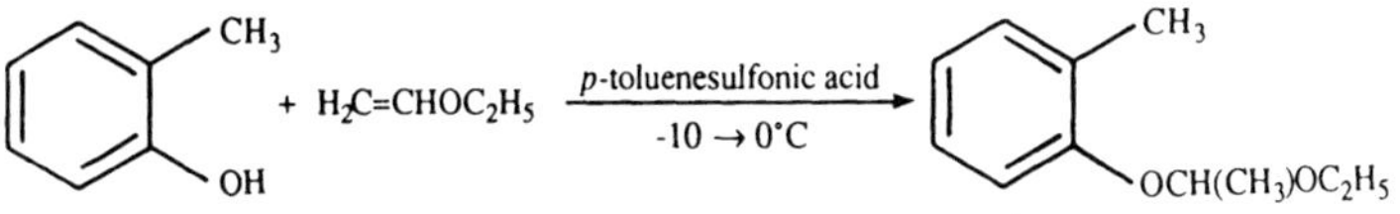

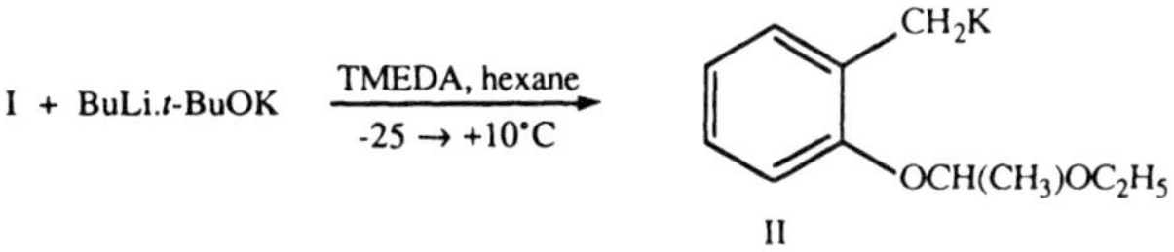

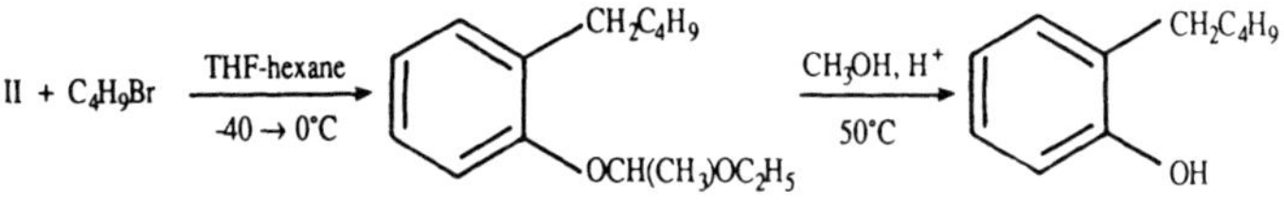

Apparatus: p. 24, 500 ml, Fig. 1, for the metallation and butylation a 1-flask is used.

Scale: 0.1 molar

Introduction

Whereas *N,N*-dimethyl-*ortho*-toluidine is specifically lithiated laterally by BuLi.TMEDA in hexane, the analogous compound *ortho*-cresyl methyl ether undergoes ring-metallation with a selectivity of ca. 90% (see Sect. 1, compound **16**) under similar conditions. Using a 1:1:1 molar mixture of BuLi, *t*-BuOK, and TMEDA in hexane ($\sim -20\,°C$), the predominant product is the α-metallated derivative ($\sim 90\,rel.\%$, 10% ring-metallation) [1]. The regioselectivity with respect to CH_3-metallation increases to at least 96%, when the metallation is carried out with the compound obtained by acid-catalyzed reaction of *p*-cresol with ethyl vinyl ether. Metallation with a 1:1 molar mixture of BuLi and *t*-BuOK in THF and hexane at temperature in the region of $-70\,°C$ gives a regioselectivity of only about 80%.

The potassium derivative **II** reacts smoothly with butyl bromide to give the alkylation product **III** in a high yield. The protecting group is readily split off under standard conditions to give *o*-pentylphenol. Other *o*-alkylphenols should also be accessible by this method. A shorter method for the preparation of *o*-alkylphenols is described in Exp. 3 (compare Ref. [2]).

Procedure

Freshly distilled ethyl vinyl ether (1.0 mol, 72.0 g) is cooled to $-10\,°C$. Ca. 300 mg of *p*-toluenesulfonic acid is added with efficient stirring (the anhydrous as well as the water-containing acid may be used). Immediately after this addition, a very concentrated solution of 0.50 mol (54.0 g) of freshly distilled *o*-cresol in a small amount of Et_2O (this solvent is merely used to liquify the cresol) is added dropwise or in small portions over about 15 min. During this addition the temperature of the solution is carefully maintained between -5 and $0\,°C$ (a cooling bath with dry ice and acetone is indispensable!). After an additional 15 min (at $0\,°C$) a second portion ($\sim 200\,mg$) of *p*-toluenesulfonic acid is added and stirring at $0-5\,°C$ is continued for 10 min (usually this second addition does not cause any heating effect). The solution is cooled to $-5\,°C$ and 10 ml of a concentrated aqueous solution of K_2CO_3 is added with vigorous stirring. After two minutes, the organic solution is dried over K_2CO_3 and subsequently concentrated in vacuo. Diethylamine or triethylamine (2 ml) is added to the remaining liquid, after which the product is distilled through a 20-cm Vigreux column (all parts of the distillation apparatus are previously rinsed with an ethereal solution of Et_2NH or Et_3N to neutralize traces of acid adhering to the glass, which may give rise to decomposition of the protected phenol). The product, b.p. $96\,°C/12$ mmHg, $n_D(20)$ 1.4880, is obtained in $\sim 95\%$ yield.

A solution of 0.12 mol of butyllithium in 78 ml of hexane is placed in the flask. Pentane or hexane (60 ml) is added, after which the solution is cooled to $-20\,°C$. Finely powdered potassium *tert*-butoxide (0.14 mol, 15.7 g, powdering should be carried out in a dry atmosphere, preferably under inert gas) is added and the mixture is stirred for 30 min at $0\,°C$. The fine suspension is then cooled to $-45\,°C$ and

TMEDA (0.20 mol, 23.2 g) is added over 2 min. The suspension is stirred for 15 min at $-40\,^\circ$C (at a moderate rate, note 1), then 0.1 mol (18.0 g) of the protected cresol is added over a few s. The temperature is allowed to rise to $-25\,^\circ$C and stirring (at the same rate) is continued for 1 h. A yellowish-brown suspension is gradually formed. The metallation is completed by removing the cooling bath and allowing the temperature to rise to $+10\,^\circ$C. THF (100 ml) is then added and the mixture is stirred for an additional 15 min at $+5\,^\circ$C (during this period the excess of BuLi $\cdot$ t-BuOK reacts with the THF to give ethene and $H_2C{=}CHOK$). After cooling the mixture (thin brown suspension) to $-60\,^\circ$C, butyl bromide (0.15 mol, 20.5 g) is added in one portion with vigorous stirring. The cooling bath is removed and the temperature may rise to $\sim -10\,^\circ$C or higher. An almost colourless (yellowish) suspension is formed. The greater part of the solvent is removed under reduced pressure (rotary evaporator). The remaining, somewhat viscous liquid is hydrolyzed with 150 ml of water, after which four extractions with pentane are carried out. The combined extracts are washed several times with a concentrated aqueous solution of ammonium chloride in order to remove the TMEDA completely (until the washings have become neutral, pH 7). After removal of the solvent in vacuo, 50 ml of methanol and 1 ml of 30% aqueous HCl are added. The solution is heated for 10 min at $\sim 50\,^\circ$C, then the methanol and the acetal $CH_3CH(OCH_3)(OC_2H_5)$ formed in the deprotection are removed under reduced pressure. Careful distillation of the remaining liquid through a 20-cm Vigreux column gives, after a small first fraction, o-pentylphenol, b.p. $120\,^\circ$C/12 mmHg, $n_D(20)$ 1.5154, in $\sim 70\%$ overall yield.

Notes

1. When stirring is carried out too vigorously, the solution may splash against the upper part of the flask, where reaction of the base with TMEDA occurs resulting in a dark deposit.

Literature

1. Heus-Kloos YA, Andringa H, Brandsma L (unpublished)
2. Bates RB, Siahaan TJ (1986) J Org Chem 51:1432

2.3 Dimetallation of *ortho*-Cresol

Functionalization reactions with butyl bromide and trimethylchlorosilane (bis-silylation).

Apparatus: p. 24, Fig. 1; 11; for the addition of the *t*-BuOK the dropping funnel is temporarily replaced by a powder funnel.

Scale. 0.1 molar (cresol)

Introduction

In a recent paper [1] Bates and Siahaan describe the OH, CH_3-dimetallation of *o*-, *m*-, and *p*-cresol with the Lochmann–Schlosser reagent and subsequent C- and/or O-functionalization of the generated intermediate with a number of electrophilic reagents. The dimetallations are carried out in refluxing alkane and take 3 (for the *ortho*- and *meta*-isomers) to 20 hours (for *para*-cresol). In order to achieve an as complete as possible conversion under these heterogeneous conditions, four (instead of two) mol equivalents of BuLi · *t*-BuOK per mol of cresol are used. In our procedure [2] the reaction is carried out under homogeneous conditions. *o*-Cresol is first added to a solution of 2-equivalents of BuLi · TMEDA in hexane to give a suspension of the lithium cresolate. Subsequent addition of *one* equivalent of *t*-BuOK results in a dissolution of the solid material. After a few hours at -20 to $-25\,°C$ and for a short period at 0 to $10\,°C$, the metallation is considered to be complete. Although any proof is lacking, we presume that in the product Li^+ is associated with O^- and K^+ with $CH_2{}^-$. Subsequent reaction with two equivalents of trimethylchlorosilane gives the expected bis-silyl compound, while alkyl bromides react specifically with the most strongly basic $CH_2{}^-$ centre to give, after acidification, the homologues of *o*-cresol. *Meta*-cresol seems to be metallated less easily and the yields of the functionalization products are lower, but nevertheless fair. Under the same conditions *p*-cresol reacts extremely sluggishly and $\sim 90\%$ of *p*-cresol iş recovered. Our findings correspond with those of Bates et al. who obtained low yields of derivatization products when electrophilic reagents were added after prolonged treatment of *p*-cresol with an excess of BuLi · *t*-BuOK under heterogeneous conditions. The successful CH_3-metallation of *o*-cresol may be explained by assuming the initial formation of a mixed aggregate from lithium *o*-cresolate and 'BuK', in which the CH_3-protons and the base are close to each other.

Procedure

To a solution of 0.21 mol of BuLi in 140 ml of hexane is added 0.21 mol (24.3 g) of TMEDA. The solution is cooled to $-50\,°C$ and a mixture of 0.1 mol (10.8 g) of *o*-cresol and 40 ml of hexane is added over a few min with cooling below $-20\,°C$. Subsequently, finely powdered (under a dry atmosphere) *t*-BuOK (0.11 mol, 12.4 g) is added over ~ 5 min while keeping the temperature between -25 and $-35\,°C$. The mixture is gently stirred for 3 h at -20 to $-25\,°C$. Most of the suspended material dissolves within 15 min and a turbid, yellow solution is formed. After the 3-h period the cooling bath is removed and the temperature allowed to rise to

10 °C. THF (120 ml) is then added, whereupon the solution is cooled to $-65\,°C$. Freshly distilled trimethylchlorosilane (0.30 mol, 32.6 g) or butyl bromide (0.12 mol, 16.4 g) is added over 10 min with vigorous stirring. After the addition, the cooling bath is removed and the temperature allowed to rise to above 0 °C (in the case of C_4H_9Br) or the mixture is warmed to 30 °C (to ensure complete O-silylation). The reaction mixture is hydrolyzed by adding ice water (200 ml) with vigorous stirring. The aqueous layer is extracted four times with Et_2O. After drying the organic solution over potassium carbonate, the solvent is removed in vacuo. Careful distillation of the remaining liquid through a 30-cm Vigreux column gives the bis-silyl derivative, b.p. 115 °C/12 mmHg in ~ 90% yield. The butyl derivative (o-pentylphenol) is liberated by acidifying the aqueous layer to pH < 2 (the extracts and the first organic layer are discarded) with 4 N aqueous hydrochloric acid. The acidic mixture is then extracted five times with a 1:1 mixture of Et_2O and pentane or hexane. The extracts are washed twice with a concentrated NaCl solution and subsequently dried over $MgSO_4$. The product is isolated in > 75% yield by distillation in vacuo (for b.p. and $n_D(20)$ see Exp. 2).

Literature

1. Bates RB, Siahaan TJ (1986) J Org Chem 51:1432
2. Andringa H, Brandsma L (unpublished)

2.4 Metallation of o- and p-Tolunitrile with Alkali Amides in Liquid Ammonia and Alkali Diisopropylamide in THF-Hexane Mixtures

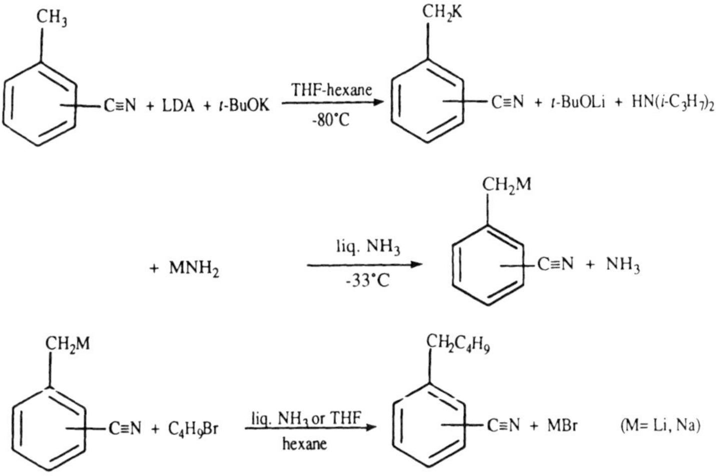

Apparatus: For the reactions in liquid ammonia a 1-1 round-bottomed, three-necked flask, equipped with a dropping funnel, a mechanical stirrer, and a gas-outlet; for the reaction in THF-hexane mixtures see p. 24, Fig. 1; 500 ml.

Scale: 0.3 or 0.1 molar for reactions in liquid ammonia or THF-hexane, respectively.

Introduction

Whereas interaction between toluene and potassium amide in liquid ammonia or lithium diisopropylamide in THF-hexane mixtures does not result in any 'ionization', addition of o- or p-tolunitrile to lithium-, sodium-, or potassium amide or LDA gives dark-coloured solutions. Addition of an alkyl bromide to the ammoniacal solutions gives rise to a vigorous reaction affording homologues of the tolunitriles in reasonable to good yields. It may be concluded that introduction of a C≡N group in the o- or p-position of CH_3 in toluene gives rise to an appreciable increase of the acidity of the methyl protons, presumably by at least three pK-units. Since in none of the resonance structures of the anion from *meta*-tolunitrile the negative charge is situated on the carbon atom bearing the C≡N group, it may be expected that the effect of a *meta*-C≡N group upon the acidity of the CH_3-protons is much smaller. The dark colour which appears when m-tolunitrile and alkali amides or alkali dialkylamides are allowed to interact, suggests the formation of *meta*-MCH_2—C_6H_4—C≡N. However, subsequent reaction with alkyl halides gives only unidentified, viscous products [1]. Moderate yield of alkylation products are obtained if the metallation of m-tolunitrile is carried out at low temperature in the presence of the co-solvent HMPT [2]. Under these conditions benzyl chloride gave only stilbene and the self-condensation product of m-tolunitrile, indicating that the deprotonation with LDA–HMPT is not complete. This self-condensation also takes place with o-tolunitrile, when this compound is treated with LDA in THF and the temperature is allowed to rise [3]. If the temperature is kept below − 70 °C, the self-condensation is slow, but under these conditions alkylations with alkyl bromides proceed sluggishly [3]. With benzophenone the expected coupling product is obtained in a reasonable yield [2], but aliphatic aldehydes and ketones give poor results [3]. Alkylations of potassiated o-tolunitrile, generated by adding o-tolunitrile to a solution of equimolar amounts of LDA and t-BuOK in THF (KDA) proceed at much lower temperatures and give the alkylated products in reasonable yields.

   In the procedure described in this experiment, o- and p-tolunitrile are metallated with alkali amides in liquid ammonia and with KDA in THF and the intermediary anions alkylated with butyl bromide. Side-reactions such as self-condensation or addition of the base across the C≡N function only take place during the reactions with *lithium* amide, but to a lesser extent than with LDA in THF.

Procedure

a) Metallation with LDA · t-BuOK in THF-hexane mixtures

A solution of 0.11 mol of butyllithium in 72 ml of hexane is placed in the flask. THF (30 ml) is added with cooling below 0 °C. The solution is then cooled to

~ − 50 °C and 0.11 mol (11.1 g) of diisopropylamine is added over a few s. Subsequently, 0.11 mol (12.3 g) of *t*-BuOK dissolved in 50 ml of THF is added with cooling below − 50 °C. The mixture is cooled (occasional cooling in a bath with liquid nitrogen) to ca. − 85 °C after which a solution of 0.1 mol (11.7 g) of *o*- or *p*-tolunitrile in 40 ml of THF is added dropwise over 15 min with cooling between − 80 and − 85 °C and stirring at a moderate rate (splashing being avoided). Butyl bromide (0.15 mol, 20.5 g) is subsequently added over 15 min to the very dark solutions, the temperature being maintained between − 80 and − 70 °C. After the addition the cooling bath is removed and the temperature allowed to rise to 0 °C. Water (200 ml) is then added to the light-brown or green solutions. After vigorous shaking or stirring, the layers are separated and the aqueous layer is extracted three times with Et_2O. The combined organic solutions are dried over $MgSO_4$, after which the solvent and other volatile compounds are removed in a water-pump vacuum. Subsequent careful distillation of the remaining liquid through a short Vigreux column gives, after small first fractions containing the starting compounds, the alkylation products, o-C_5H_{11}—C_6H_4C≡N, b.p. ~ 86 °C/1 mmHg, $n_D(20)$ 1.5094, and p-C_5H_{11}—C_6H_4—C≡N, b.p. 103 °C/1 mmHg, $n_D(20)$ 1.5118 in 64 and 67% yields, respectively.

b) Metallation with alkali amides in liquid ammonia

Suspensions of lithium amide and sodamide in ~ 400 ml of liquid ammonia are prepared from 0.35 mol (2.5 g) of lithium and 0.36 mol (8.3 g) of sodium (see Vol. 1, Chap. I). A mixture of 0.3 mol (35.1 g) of *o*- or *p*-tolunitrile and 30 ml of THF is added dropwise over 7 min with vigorous stirring. After an additional 5 min, butyl bromide (0.45 mol, 62 g) is added over 7 min to the dark-red solutions. The reactions are very vigorous. After an additional 15 min, the remaining ammonia is evaporated by placing the flask in a water bath at 40 °C (the outlet and dropping funnel are removed). Water (300 ml) is added to the remaining salty mass and dissolution of the salts is effected by vigorous stirring. The products are isolated as described above (extraction with dichloromethane may be more convenient than with Et_2O). The yields of the products from *o*- and *p*-tolunitrile are 66 and 67% in the case of metallation with lithium amide. The reaction with *o*-tolunitrile and $LiNH_2$ gives a larger amount (~ 8 g) of (viscous) residue, while in the case of *p*-tolunitrile there is more of the first fraction (incomplete conversion). With sodamide the yields are ~ 10 and 5%, respectively, higher.

Literature

1. Rash FH, Boatman S, Hauser CR (1967) J Org Chem 32:372
2. Kaiser EM, Petty JD (1976) J Organometal Chem 107:219
3. Tip L, Brandsma L (unpublished)

2.5 α-Lithiation of *p*-Toluene-*N*,*N*-Dimethylsulfonamide

Functionalization with trimethylchlorosilane.

Apparatus: p. 24, Fig. 1; 500 ml

Scale: 0.05 molar

Introduction (compare Exp. 4)

Like the C≡N substituent in the preceding experiment, the SO_2NMe_2-group has strong electron-withdrawing properties, which should facilitate the removal of a proton from a methyl group, especially when it is in the *ortho*- or *para*-position. It is shown in this experiment that *p*-toluene-*N*,*N*-dimethylsulfonamide is readily lithiated by LDA in a mixture of THF and hexane. Treatment of the reaction mixture with trimethylchlorosilane gives the α-silylated derivative in a high yield. There are no indications for the occurrence of *ortho*-(ring)metallation. Even the less stronger bases lithium amide and sodamide in liquid ammonia (used in excess) are capable of effecting appreciable ionization as appears from the good yields of homologues formed by subsequent addition of alkyl bromides. It is worth mentioning in this connection that with *butyllithium* in Et_2O or THF a very rapid and exclusive metallation at the *ortho*-position of the SO_2NMe_2 group takes place. In Et_2O this white suspension is very unreactive (probably as a consequence of internal coordination) but in THF it reacts with trimethylchlorosilane to give the expected trimethylsilyl derivative as a viscous liquid.

Procedure

A mixture of 0.06 mol (6.1 g) of diisopropylamine and 50 ml of THF is cooled to below − 40 °C, after which a solution of 0.06 mol of butyllithium in 40 ml of hexane is introduced over a few min from a syringe, while keeping the temperature below 0 °C. After cooling the resulting solution of LDA to − 40 °C, a solution of 0.05 mol (10.0 g) of the sulfonamide (see below) in 40 ml of THF is added over 5 min with cooling between − 30 and − 40 °C. After the addition, the cooling bath is removed and the temperature allowed to rise to 0 °C. The thick yellow suspension is cooled again to − 40 °C and freshly distilled trimethylchlorosilane (0.07 mol, 7.6 g) is added in one portion with vigorous stirring. The cooling bath is removed and the temperature allowed to rise to 0 °C. Water (100 ml) is then added with vigorous stirring. The organic layer together with one ethereal extract are dried over $MgSO_4$ and subsequently concentrated in vacuo, the last traces of volatile compounds being removed in a vacuum of < 1 mmHg. The remaining solid is dissolved in refluxing Et_2O ($\sim$ 70 ml), pentane or hexane ($\sim$ 200 ml) is added and the solution cooled to − 30 °C to give crystals (long needles) of the α-silylated sulfonamide, m.p. 108.6 °C (Mettler FP5). Concentration of the mother liquor and cooling to − 30 °C gives an additional small amount of crystalline material (pure according to ^1H NMR), bringing the yield at $\sim$ 85%.

Preparation of p-toluene-N,N-dimethylsulfonamide

Liquified (− 50 °C) dimethylamine (0.50 mol, 22.5 g) is poured into 200 ml of dichloromethane cooled to − 50 °C. A mixture of 0.2 mol (38.1 g) of p-toluenesulfonyl chloride and 150 ml of dichloromethane is added dropwise with cooling between − 50 and − 30 °C. After the addition the cooling bath is removed and the temperature allowed to rise to + 10 °C. A thin suspension is formed. Water (200 ml) is added with vigorous stirring. The organic layer is washed once with water and subsequently dried over $MgSO_4$. Removal of the solvent in vacuo (the last traces at < 1 mmHg) gives the sulfonamide in an almost quantitative yield.

Chapter V
Metallation of Heterosubstituted Allylic and Benzylic Compounds

1 Introduction

An extensive review [1] covers the metallations of a wide variety of allylic compounds C=C—C—X and benzylic compounds Ar—CH—X in which X represents a heterosubstituent having the heteroatom directly linked to the allylic or benzylic system. The experimental part of this chapter is confined to procedures for the metallation of simple allylic and benzylic ethers, silanes, and sulfides. The experiments illustrate the use of a variety of strong bases as deprotonating reagents. For more complicated substrates containing substituents that stabilize negative charge, the conditions for metallation in many cases can be easily derived. Table 6 gives a survey of metallations carried out under 'preparative' conditions which are described in the experimental section. The list is extended with some data taken from the review [1] (indicated with B.D.).

Literature

1. Biellmann JF, Ducep JB (1982) Org Reactions 27:1, John Wiley

2 Substrates and Metallation Conditions

2.1 Allylic Amines and Ethers

There has been some discussion (see Org Reactions 27:4) about the relative thermodynamic acidities of ethers and the corresponding compounds in which O is replaced by hydrogen. It should in any case be clear from our preparative experiments that allylic ethers, $H_2C=CHCH_2OR$ $(R=CH_3)$, and allylic amines, $H_2C=CHCH_2NR_2$ $(R=CH_3)$, are metallated more easily than aliphatic 1-alkenes, $H_2C=CHCH_2R$. Methyl allyl ether, for example, can be lithiated at low temperatures ($\sim -30\,°C$) with BuLi in a mixture of THF and hexane. For the metallation of propene and homologues, the super-basic reagent BuLi·t-BuOK is required and good results can be obtained only when the alkene is used in a large excess (compare Chap. II, Table 3 with this Chapter, Table 6). The conditions for the metallation of $H_2C=C(CH_3)OCH_3$, which is a *heterosubstituted* propene rather than an *allylic ether*, are closely similar to those for propene (see Chap. II).

Table 6. Representative conditions for the generation of heterosubstituted allylic and benzylic anions[a]

Substrate (mol% excess)	Base–solvent system (mol% excess)	Temp. range[b] of metallation in °C (time in min)	
$H_2C=CHCH_2NMe_2$[c] (30)	BuLi·TMEDA, hexane	20(0) → 50(10)	
$H_2C=CHCH_2NEt_2$ (20)	BuLi·t-BuOK, THF-hexane	−75(10) → −55(10)	
$H_2C=CHCH_2N(CH_3)Ph$	BuLi·t-BuOK, alkane	0(30) → 20(30)	B, D
$H_2C=CHCH_2NHC(=O)t\text{-Bu}$	2LDA, $CH_3OCH_2CH_2OCH_3$	−78	B, D
$H_2C=CHCH_2OCH_3$[c,d] (30)	BuLi, THF-hexane	−60(0) → −40(40)[f]	
(20)	BuLi·t-BuOK, THF-hexane	−90(10) → −70(5)	
$H_2C=CHCH_2Ot\text{-Bu}$[c,d] (30)	BuLi·TMEDA, hexane	−40(10) → −35(90)[d]	
(20)	BuLi·t-BuOK, THF-hexane	−80(10) → −60(5)	
$H_2C=CHCH_2OR^d$ (R = −CH(CH_3)OEt) (10)	BuLi·t-BuOK, THF-hexane	−90(10) → −70(15)[f]	
$H_2C=C(CH_3)OCH_3$ (300)	BuLi·t-BuOK, THF-hexane	−75(30) → −50(30)	
$H_2C=CHCH_2SCH_3$ (10)	BuLi, THF-hexane	−80(10) → −20(0)	
	KNH_2, liq. NH_3	−33(0)[e,f]	
$H_2C=CHCH_2SPh$	BuLi, THF-hexane (5)	−80(10) → −20(0)	
	KNH_2, liq. NH_3(10)	−33(0)[f]	
$(CH_3)_2C=CHCH_2SPh$	BuLi, THF	−30(30)	B, D
$H_2C=CH$—[1,3-dithiane]	BuLi, THF-hexane	−78 → 0	B, D
$H_2C=CHCH$=[1,3-dithiane]	BuLi, THF-hexane	−80 (1,4-addition)	B, D
$(CH_3)_2C$=[1,3-dithiane]	BuLi, THF-hexane, HMPT	−78	B, D
$H_2C=CHCH=CHSCH_3$	BuLi, Et_2O-hexane (10)	−20(0) → 15(30)[f] (1,4-addition)	
CH_3SCH_2CH=[1,3-dithiane]	LDA, THF-hexane, HMPT	−78	B, D
C_4H_9CH=[1,3-dithiane]	LDA, THF-hexane, HMPT	−78	B, D
$(CH_3)_2C=CH$—[1,3-dithiane]	BuLi, THF-hexane	−20(360)	B, D
H—[cyclohexanylidene]=[1,3-dithiane]	BuLi, THF-hexane	−80	B, D
[cyclohexadienyl, H]=[1,3-dithiane]	LDA, THF-hexane, HMPT	−78	B, D

(Continued)

Table 6. (*Continued*)

Substrate (mol% excess)	Base–solvent system (mol% excess)	Temp. range[b] of metallation in °C (time in min)	
$H_2C=CHCH_2SH$	2 BuLi, TMEDA, hexane	0(180) (dimetallation)	B, D
	2 BuLi, TMEDA, THF-hexane	0(240) (dimetallation)	B, D
$PhCH_2SH$	2 BuLi; TMEDA, pentane	-5 (150) (dimetallation)	B, D
$H_2C=CHCH_2S(=O)Ph$	BuLi, THF-hexane	-50	B, D
	LDA, THF-hexane	$-60(15)$	B, D
	$LiNH_2$, liq. NH_3	$-33(0)^f$	
$CH_3CH=CHCH_2S(=O)Ph$	LDA, THF-hexane	$-60(15)$	B, D
$PhCH_2SCH_3$	BuLi, THF-hexane (5)	$-50(0) \rightarrow -20(10)$	
$H_2C=CHCH_2SiMe_3$ (10)	BuLi, TMEDA, hexane	$10(0) \rightarrow 30(30)$	
$PhCH_2SiMe_3$	BuLi, TMEDA, hexane (5)	$+20(0) \rightarrow 45(10)$	
$PhCH_2NMe_2$	BuLi·t-BuOK, THF-hexane (10)	$-90(0) \rightarrow -75(45)$	
$H_2C=CHCH_2SePh$	LDA, THF-hexane	$-78(10)$	B, D
$PhSeCH_2Ph$	LDA, THF-hexane	-30	B, D

[a] Taken from our own experiments and from the tables in the review of Biellmann and Ducep (indicated as B, D in the last column), see Sect. 1, Ref. [1]. In our procedures initial concentrations of substrate and base are of the order of 0.5 to 0.8 mol/l; the concentrations in other experiments (B, D) may be lower.

[b] A denotation like $-90(10) \rightarrow -60(30)$ means: after a quick addition at -90 °C of the substrate to the base, the reaction mixture is kept for 10 min at -90 °C, then the temperature is allowed to rise (without cooling) to -60 °C (or brought at -60 °C by some warming). In order to ensure a complete metallation the reaction mixture is stirred for an additional 30 min at -60 °C.

[c] The methyl ethers and dimethylamines are metallated more easily than the t-butyl ethers and diethylamines, respectively.

[d] Above -10 °C a Wittig-rearrangement sets in (compare Felkin H, Tambuté A (1969) Tetrahedron Lett 821). Even at -20 °C to -35 °C Wittig rearrangement (1,2- and 1,4-) occurs to some extent.

[e] The metallation may be incomplete under these conditions.

[f] Unpublished experimental data, not described in the experimental section.

Whereas lithiated N,N-dialkyl allylamines, $H_2C=CHCH(Li)NR_2$, have a reasonable thermal stability, allowing their generation at reflux temperature, the analogous metallated ethers undergo Wittig-rearrangement at temperatures above -20 °C. Using the super-basic reagent BuLi·t-BuOK, a smooth potassiation can be effected in THF at temperatures between -60 and -90 °C. Wittig-rearrangement (in the case of allylic ethers) does not occur under these conditions. Also for the metallation of allylic amines, we found BuLi·t-BuOK to be more suitable than other reagents such as BuLi·TMEDA.

2.2 Allylic and Pentadienylic Sulfur, Silicon, and Selenium Compounds

a) Metallation

A smooth metallation of allylic sulfides can be generally achieved at low temperatures with n-butyllithium in a mixture of THF and hexane. The use of the more strongly basic *sec*-butyllithium has no special advantages. Although alkali amides (especially $NaNH_2$ and KNH_2) are sufficiently strong bases to effect complete metallation of allylic S,S-acetals ($C=C-CH(SR)_2$), allylic sulfoxides ($C=C-CH-SOR$), and sulfones ($C=C-CH-SO_2R$), they have been applied in incidental cases only. The main reason may be that liquid ammonia is not a very popular solvent: many chemists associate working with this solvent with a complicated apparatus, consisting *inter alia* of a reflux condenser cooled with dry ice and acetone. Some of them even prefer the combination of LDA with the poisonous (carcinogenic!) co-solvent HMPT over that of $NaNH_2$ and liquid ammonia. We wish to point out, that liquid ammonia is an excellent solvent for alkylations with alkyl bromides and β-hydroxyalkylations with epoxides. Metallated allylic sulfoxides and (especially) sulfones react sluggishly in the 'usual' solvent THF with unactivated alkyl halides, but in the polar liquid ammonia (at its b.p. $-33\,°C$) the alkylations are fast[1].

The group R in allylic sulfides, $H_2C=CHCH_2SR$, and in the isomers, $CH_3CH=CHSR$, has a marked influence upon the position of the metallation equilibria in liquid ammonia [1]:

$$H_2C=CHCH_2SR + MNH_2 \rightleftharpoons H_2C=CH-CH-SR + NH_3$$
$$\qquad\qquad\qquad\qquad\qquad\qquad\qquad\underset{\displaystyle M}{\vert}$$
$$\rightleftharpoons CH_3CH=CHSR + MNH_2$$
$$(M = Li, Na, K)$$

Upon addition of $H_2C=CHCH_2SCH_3$ to a suspension of lithium amide in liquid ammonia only a faint yellow colour appears. In the cases of $H_2C=CHCH_2SCH=CH_2$ and $H_2C=CHCH_2SPh$, the suspension initially disappears and yellow to brown solutions of the anion appear. Immediate addition of an alkyl bromide gives the alkylation products in excellent yields. If the solutions of the anions are allowed to stand for more than 15 min, the suspension of $LiNH_2$ reappears due to partial ammonolysis of the anions with formation of the propenylic sulfides, $CH_3CH=CHSR$. Interaction between sodamide or potassium amide and either $H_2C=CHCH_2SCH=CH_2$ ($-SPh$) or the propenylic isomers presumably gives rise to a much more complete formation of the anions. When R = alkyl, however, only potassium amide is capable of generating the allylic anions in significant concentrations. Pentadienylic sulfides, such as $H_2C=CHCH=CHCH_2SCH_3$[1] and 2H-thiopyran [2], are more acidic (pK difference presumably 1 to 2 units) than allylic or propenylic sulfides and high equilibrium concentrations of the anions are produced in liquid ammonia with $LiNH_2$, $NaNH_2$, and KNH_2. With the latter base, the 'ionizations' are probably complete. Introduction of an additional SR substituent in the allylic systems also gives rise to a lowering of the pK. Thus, 'ionization' of the S,S-acetals $H_2C=CHCH(SR)_2$ and $CH_3CH=C(SR)_2$ (R = alkyl) under the influence of $NaNH_2$ or KNH_2 is presumed to be (almost) complete [1].

---

[1] For working with liquid ammonia see our previous books Brandsma L, Verkruijsse HD (1987) Preparative Polar Organometallic Chemistry, Vol. 1, Springer-Verlag; and Brandsma L (1988) Preparative Acetylenic Chemistry, Revised Edition, Elsevier

Allylic sulfoxides and sulfones are completely metallated in liquid ammonia by all alkali amides [1]. For metallation in organic solvents, LDA and BuLi are usually applied. The lithiation of allyltrimethylsilane can be conveniently achieved with BuLi·TMEDA (or HMPT) or t-BuLi·TMEDA (or HMPT) in THF-alkane mixtures [3]. As in the case of the sulfur compounds, extension of the unsaturated system leads to an increased acidity. Thus H_2C=CHCH=CHCH$_2$SiMe$_3$ can be metallated with the less strongly basic LDA in THF [4]. For the metallation of allylic selenides, LDA seems to be the reagent of choice. Butyllithium will presumably attack on selenium.

b) Addition

Strongly basic reagents such as butyllithium add to the diene system of certain conjugated diene sulfides and keten-S,S-acetals to give allylic anions [5]:

$$H_2C=CH-CH=C(R)SR' + BuLi \longrightarrow BuCH_2CH=CH-C(Li)(R)SR'$$
$$(R=H \text{ or } SR')$$

Literature

1. Unpublished observations in the laboratory of the author
2. Gräfing R, Brandsma L (1980) Recl Trav Chim Pays-Bas 99:23
3. Lau PWK, Chan TH (1978) Tetrahedron Lett 2383
4. Kauffmann TT, Gaydoul KR (1985) Tetrahedron Lett 26:4067
5. Seebach D, Kolb M, Gröbel BT (1973) Angew Chemie 85:42, 189; (1973) Angew Chemie Int ed. (Engl) 12:69

2.3 Benzylic Amines, Silanes, and Sulfides

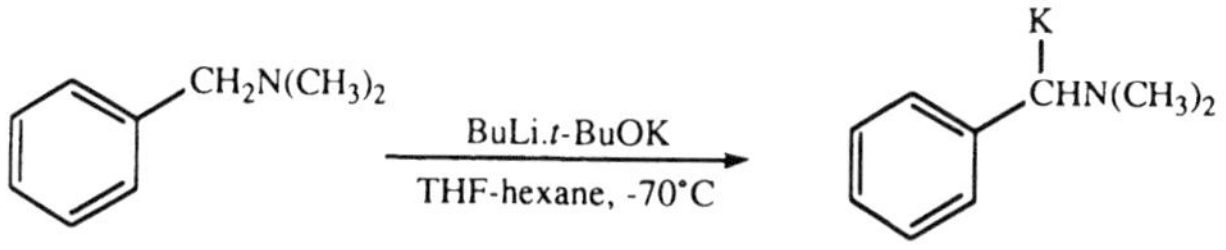

The classical example of a coordination-induced *ortho*-lithiation is the reaction of PhCH$_2$N(CH$_3$)$_2$ with butyllithium [1], giving o-Li—C$_6$H$_4$—CH$_2$N(CH$_3$)$_2$. With BuLi·t-BuOK in a mixture of THF and hexane the only reaction is α-metallation [2;3], while BuLi·TMEDA in hexane gives rise to both ring- and α-lithiation [2].

Treatment of methyl benzyl sulfide with BuLi in THF results in a smooth and specific lithiation at the methylene carbon atom. Even potassium amide in liquid ammonia is capable of effecting metallation [2]. The second lithiation of benzyl mercaptan with BuLi·TMEDA occurs exclusively at the methylene carbon atom [4]:

$$PhCH_2SCH_3 \xrightarrow[\text{THF-hexane}]{BuLi} PhCH(Li)SCH_3$$

$$PhCH_2SH \xrightarrow[\text{hexane}]{2 \text{ BuLi·TMEDA}} PhCH(Li)SLi$$

A trimethylsilyl group is a less strongly acidifying substituent than an alkylthio group. Thus, whereas $PhCH_2SCH_3$ is readily lithiated at low temperatures by butyllithium in THF, the silicon analogue $PhCH_2SiMe_3$ reacts sluggishly under these conditions. Lithiation at room temperature cannot be achieved without serious competitive attack of THF by BuLi. Quantitative metallation is effected by the BuLi·TMEDA reagent [2, compare 5]:

$$PhCH_2SiMe_3 \xrightarrow[\text{room temperature}]{\text{BuLi·TMEDA}} PhCH(Li)SiMe_3$$

Literature

1. Gschwend HW, Rodriguez G (1979) Org Reactions 26:1, John Wiley
2. Brandsma L (unpublished observation)
3. Puterbaugh WH, Hauser CR (1963) J Am Chem Soc 85:2467
4. Geiss KH, Seebach D, Seuring B (1977) Chem Ber 110:1833
5. Peterson DJ (1968) J Org Chem 33:780

2.4 Regiochemistry of the Reactions of Allylic Alkali Metal Compounds with Electrophiles

An allylic metal compound with the system $X-\overset{\alpha}{C}-C-\overset{\gamma}{C}-$ has two carbon atoms $C\alpha$ and $C\gamma$ available for attack by an electrophilic reagent:

$$X-C-C-C \xrightarrow{E} X-\overset{\overset{\displaystyle E}{|}}{C}-C{=}C + X-C{=}C-C-E$$

$$\text{M} \qquad\qquad \alpha\text{-attack} \qquad \gamma\text{-attack}$$

M = Li, Na, K; X = alkyl, aryl or a heterosubstituent, mostly R_2N, RO, RSe, RS, R_3Si, RS(=O)-, RSO_2-, $R-C{=}O$, COOR, $C{\equiv}N$.

Depending upon the nature of the alkali-metal ion, the solvent, the group X and other substituents the ratio of the products derived from α- and γ-attack shows greater or smaller variations. Since separation of the two types of products is generally difficult, only reactions that give exclusively or predominantly one type of product have preparative importance.

Gompper and Wagner [1] have made an attempt to correlate the regiochemistry with the electron distribution (caused by the substituents, in particular those with electron-donating and -accepting properties) which can be approached calculationally. The organic chemist who considers to apply some allylic metal compound in a synthesis, usually wants to make a quicker prediction of the outcome, however.

In their outstanding review [2], Biellmann and Ducep have tabulated a large number of reactions of hetero-substituted allylic 'anions'. Considerations about the regiochemistry are given on pages 35–40 of this review, on pages 657 and 658 of the review by Martin [3], and in the review of Werstiuk [18].

Based on the various literature data and our own experiments (see also Chap. II) some conclusions can be drawn:

1. The high preference of Me_3SiCl and $FB(OCH_3)_2$ for attack on C-γ is rather general, permitting the synthesis of pure silanes $XCH=CHCH_2SiMe_3$ (X=alkyl, see, for example, Ref. [4]) and boron compounds $XCH=CHCH_2B(OCH_3)_2$ (X=alkyl). The latter can be converted into the corresponding alcohols alkyl-$CH=CHCH_2OH$ by treatment with hydrogen peroxide [5].

2. Dialkyl disulfides attack lithiated allylic ethers with high ($> 90\%$) selectivity at the γ-position [6].

3. Lithiated (and also potassiated) allylic ethers and tertiary amines undergo mainly γ-attack by primary alkyl halides. The preference for C-γ is particularly strong with the metallated allylic ethers $H_2C=CHCH_2Ot$-Bu and $H_2C=CHCH_2OSiEt_3$ [6, 7]. Selectivities in reactions with sec-alkyl halides, allylic and cyclohexyl halides are often considerably lower than with primary alkyl halides [6, 8]. Metallated allylic sulfides show mainly α-attack by primary alkyl halides [2, 8, 17].

4. Allylic anions having a strong electron-withdrawing group on the α-C atom are alkylated exclusively on C-α. Examples are $H_2C=CH-\bar{C}H-C\equiv N$ and $H_2C=CH-\bar{C}H-COOR$ (see Chaps. VIII and X).

5. The regioselectivity with simply-substituted epoxides is often parallel with that of primary alkyl halides.

6. The presence of an anionic group next to the allylic anion causes γ-selectivity for all kinds of electrophiles. Examples are the di-anions $H_2\bar{C}-CH=CHS^-$ [9], t-Bu$-C(=O)\bar{N}-CH=CHCH_2$ [10], $H_2\bar{C}-CH=CH-CH=C(SR)S^-$ [11] and R$-CO\bar{C}H-CH=CHCH_2$ [12].

7. Lithiated allyl silanes $LiCH_2CH=CHSiMe_3$ show a useful γ-regio-preference in reactions with carbonyl compounds [13]. With the lithiated N,O and S-analogues this regiochemistry is very variable (2, 6, 8], sometimes totally unexpected and not very useful from a preparative point of view.

8. Alkylation of cyclic or non-cyclic pentadienylic anions with primary alkyl halides gives mainly or exclusively the non-conjugated dienes. Thus, reaction of the lithium compound from 1,4-cyclohexadiene with butyl bromide gives the non-conjugated and conjugated butyl derivatives in a 3:1 ratio [14]. The main product from $CH_3CH=CHCH=CH_2$, potassium amide in liquid ammonia and butyl bromide is $H_2C=CHCH(C_4H_9)CH=CH_2$ [8]. Alkylation of the anion from 2H-thiopyran with primary alkyl halides in organic solvents and in liquid ammonia gives more than 95% of the deconjugated alkyl derivatives [8]. Reaction of $H_2C=CHCH=CHCH-SR$ with primary alkyl halides results mainly in the formation of the α-product $H_2C=CHCH=CHCH(R')SR$ [8], while ketones react exclusively at C-α [15]. Trimethylsilylation of $H_2C=CHCH=CHCH_2Li$ proceeds with perfect selectivity to give $H_2C=CHCH=CHCH_2SiMe_3$ [16].

Literature

1. Gompper R, Wagner HU (1976) Angew Chemie 88:389; (1976) Angew Chem Int ed (Engl) 15:437
2. Biellmann JF, Ducep JB (1982) Allylic and benzylic carbanions substituted by heteroatoms. Org Reactions 27:1, John Wiley
3. Martin SF (1979) Synthesis 633
4. Mordini A, Palio G, Ricci A, Taddei M (1988) Tetrahedron Lett 4991
5. Rauchschwalbe G, Schlosser M (1975) Helv Chim Acta 58:1094
6. Evans DA, Andrews GC, Buckwalter B (1974) J Am Chem Soc 96:5560
7. Clark Still W, Macdonald TL (1974) J Am Chem Soc 96:5561
8. Unpublished observations made in the course of the experimental preparation of this book.
9. Geiss KH, Seebach D, Seuring B (1977) Chem Ber 110:1833
10. Tischler AN, Tischler MH (1978) Tetrahedron Lett 3
11. Pohmakotr M, Seebach D (1979) Tetrahedron Lett 2271; compare Gräfing R, Brandsma L (1979) Recl Trav Chim Pays-Bas 98:520
12. Seebach D, Pohmakotr M, Schregenberger C, Weidmann B, Mali RS, Pohmakotr S (1982) Helv Chim Acta 65:419
13. Ayalon-Chass D, Ehlinger E, Magnus P (1977) J Chem Soc, Chem Comm 772
14. Brieger G, Anderson DW (1970) J Chem Soc Chem Comm 1325
15. Kauffmann T, Gaydoul KR (1985) Tetrahedron Lett 4071
16. Seyferth D, Pornet J (1980) J Org Chem 45:1721; Oppolzer W, Burford SC, Marazza F (1980) Helv Chim Acta 63:555
17. Gröbel BT, Seebach D (1977) Synthesis 392
18. Werstiuk NH (1987) in: Hase TA (ed) Umpoled synthons, John Wiley, p 173

3 Experiments

All temperatures are internal, unless indicated otherwise.

For general instructions concerning handling alkali-metal reagents, drying solvents, etc., see Vol. 1, Chap. I.

All reactions are carried out in an atmosphere of inert gas, except those in boiling ammonia.

3.1 Metallation of *N,N*-Dialkyl Allylamines with BuLi·TMEDA and BuLi·*t*-BuOK

$$H_2C\!=\!CHCH_2NMe_2 + BuLi \cdot TMEDA \xrightarrow[20 \to 50\,^{\circ}C]{\text{hexane}}$$

$$H_2C\!=\!CHCH(Li)NMe_2 + C_4H_{10}$$

$$H_2C\!=\!CHCH_2NEt_2 + BuLi \cdot t\text{-BuOK} \xrightarrow[-80 \to -30\,^{\circ}C]{\text{THF-hexane}}$$

$$H_2C\!=\!CHCH(K)NEt_2 + t\text{-BuOLi}$$

$$H_2C = CHCH(Li)NMe_2 + C_5H_{11}Br \xrightarrow[< -30\,°C]{THF\text{-}hexane}$$

$$C_5H_{11}CH_2CH = CHNMe_2 + LiBr$$

$$H_2C = CHCH(K)NEt_2 + Me_3SiCl \xrightarrow[< -30\,°C]{hexane}$$

$$Me_3SiCH_2CH = CHNEt_2 + KCl$$

Apparatus: p. 24; Fig. 1; 500 ml

Scale: 0.1 molar

Introduction

There are only a few reports on the metallation of allylic amines, $H_2C = CHCH_2NR_2$. The amine in which NR_2 represents a pyrrolidino group has been lithiated with an excess of *sec*-BuLi in a mixture of THF and hexane [1], while the metallation of $H_2C = CHCH_2N(CH_3)Ph$ has been achieved in petroleum ether with a 1:1 molar mixture of BuLi and *t*-BuOK [2]. Lithiation of *N*-allylcarbazole is effected at $-15\,°C$ by *n*-BuLi · TMEDA in Et_2O [3], which indicates that aryl groups linked to nitrogen make the lithiation more facile. We found that the metallation of *N,N*-dimethyl- and *N,N*-diethyl allylamine with BuLi in THF-hexane mixtures proceeds inefficiently [4] (slow reaction below $0\,°C$, serious competitive attack of THF at room temperature). Excellent results are obtained with a 1:1 molar mixture of BuLi and *t*-BuOK in THF and hexane, while BuLi · TMEDA is a capable of effecting a smooth lithiation of $H_2C = CHCH_2N(CH_3)_2$ at temperatures in the region of $40\,°C$. The allylic *diethyl*amine reacts much more slowly and competitive metallation of TMEDA [5] cannot be avoided unless a large excess of the allylic amine is used. In this experiment the metallation of the two allylic amines with BuLi·*t*-BuOK and BuLi·TMEDA and the reactions of the allylic metal compounds with *n*-hexyl bromide and trimethylchlorosilane are described. In both cases the predominant product is the γ-functionalized derivative. The preferential *Z*-configuration of the allylic anions is retained during the functionalizations. It should be pointed out, however, that it is absolutely necessary to carry out the work-up under strictly neutral (or weakly basic) conditions. Traces of acid adhering to the glass can cause a very ready isomerization to the *E*-isomers. This isomerization may also take place in the NMR tube under the influence of acid present in $CDCl_3$ or CCl_4. All glassware (distillation apparatus and NMR tube) should be rinsed with gaseous ammonia or a dilute solution of diethylamine in acetone, and subsequently blown dry.

Procedure

Metallation of $H_2C = CHCH_2N(CH_3)_2$ with BuLi · TMEDA in hexane

A solution of 0.1 mol of butyllithium in ~ 67 ml of hexane is placed in the flask. TMEDA (0.1 mol, 11.6 g) is added in one portion at room temperature. After cooling to ca. 25 °C, 0.13 mol (11.1 g) of N,N-dimethyl allylamine (Exp. 8) is added in one portion. The temperature may rise to between 35 and 40 °C over a few min. The gently stirred yellow to orange solution is warmed for an additional 10 min at 50 °C, then it is cooled to 0 °C and 70 ml of THF is added. The solution is further cooled to − 50 °C, after which n-pentyl bromide (0.1 mol, 15.1 g) is added over 10 min with vigorous stirring. After the addition, the cooling bath is removed and the temperature allowed to rise to 0 °C. Water (100 ml) is then added with vigorous stirring. The organic layer and three ethereal extracts are dried over potassium carbonate. The organic solution is concentrated under reduced pressure and the remaining liquid carefully distilled through a 40-cm Vigreux column. The product, b.p. 75 °C/12 mmHg, $n_D(20)$ 1.4482 (yield ∼ 75%), consists for ⩾ 95% of Z-$C_5H_{11}CH_2CH=CHN(CH_3)_2$. Cyclohexyl bromide gives a mixture of comparable amounts of α- and γ-alkylation product in ∼ 60% yield.

Metallation of N,N-diethyl allylamine with BuLi·t-BuOK in THF

THF (40 ml) and N,N-diethyl allylamine (0.12 mol, 13.6 g, see Exp. 8) are placed in the flask. A solution of 0.1 mol of BuLi in 67 ml of hexane is added with cooling below 0 °C. The mixture is cooled to − 80 °C (occasional cooling in a bath with liquid N_2), after which a solution of 0.1 mol (11.2 g) of t-BuOK in 50 ml of THF (if the quality of t-BuOK is good, the solution is almost clear) is added over 10 min with stirring at a moderate rate. The temperature of the orange solution is maintained between − 80 and − 70 °C. After the addition the temperature is allowed to rise to − 30 °C and the solution is kept at that level for an additional 10 min, then a mixture of 0.11 mol (11.9 g) of freshly distilled trimethylchlorosilane and 40 ml of Et_2O is added over a few s with vigorous stirring. Ice water (100 ml) is added (also with vigorous stirring) to the almost colourless gelatinous suspension. The organic solution is washed four times with ice water, the washings being combined with the first aqueous layer. The combined aqueous layers are extracted twice with a 1:1 mixture of Et_2O and pentane. The extracts are washed once with ice water and then combined with the other organic solution. After drying over potassium carbonate ($MgSO_4$ might cause Z- to E-isomerization), the solvent is removed under reduced pressure. Careful distillation of the remaining liquid through a 40-cm Vigreux column gives the product (⩾ 95% Z-$Me_3SiCH_2CH=CHN(C_2H_5)_2$), b.p. 70 °C/12 mmHg, $n_D(20)$ 1.4480, in greater than 80% yield.

Literature

1. Martin SF, DuPriest MT (1977) Tetrahedron Lett 3925
2. Ahlbrecht H, Eichler J (1974) Synthesis 672; Ahlbrecht H, Vonderheid C (1975) Synthesis 512
3. Julia M, Schouteeten A, Baillargé M (1974) Tetrahedron Lett 3433
4. Andringa H, Brandsma L (unpublished observations)
5. Köhler FH, Hertkorn N, Blümel J (1987) Chem Ber 120:2081

3.2 Metallation of Allyl *t*-Butyl Ether with BuLi·*t*-BuOK

$$H_2C=CHCH_2Ot\text{-}Bu + BuLi \cdot t\text{-}BuOK \xrightarrow[-80 \to -60°C]{THF\text{-}hexane}$$

$$H_2C=CHCH(K)Ot\text{-}Bu + t\text{-}BuOLi + C_4H_{10}$$

$$H_2C=CHCH(K)Ot\text{-}Bu + C_6H_{13}Br \longrightarrow C_6H_{13}CH_2CH=CHOt\text{-}Bu$$
$$\text{(main product)}$$

Apparatus: p. 24; Fig. 1; 500 ml

Scale: 0.1 molar

Introduction

Allylic ethers are metallated more easily than the analogous amines. Whereas the lithiation of alkyl-OCH$_2$CH=CH$_2$ with BuLi·TMEDA in hexane or BuLi in THF-hexane mixtures can be completed within 30 to 90 min at temperatures between -30 and $-40°C$, the amines H$_2$C=CHCH$_2$N(CH$_3$)$_2$ and H$_2$C= CHCH$_2$N(C$_2$H$_5$)$_2$ are hardly attacked by BuLi under these conditions. Evans and Still et al. used the more strongly basic *sec*-BuLi in THF-hexane mixtures for the metallation of allylic ethers [1,2]. With this reagent a rapid metallation can be achieved at $-65°C$. At these low temperatures no Wittig-rearrangement [1,3] of the allylic anions occurs. *n*-Butyllithium reacts more slowly and higher temperatures are required. The limit lies around $-35°C$; above this temperature the metallated ethers begin to rearrange.

$$H_2C=CH-\underset{\sim}{C}H-OR \xrightarrow[\text{(1,2- and 1,4-)}]{Wittig} H_2C=CH-CH(R)-O^-$$
$$+ R\underset{Z}{CH_2CH=CH}-O^-$$

The ratio of 1,2- and 1,4-Wittig products (determined after silylation) depends *inter alia* upon the group R. A good alternative for *sec*-BuLi is the 1:1 molar mixture of *n*-BuLi and *t*-BuOK which is completely soluble in THF (see Vol. 1, Chap. I). It may be prepared by mixing BuLi in hexane with *t*-BuOK in THF at temperatures below $-70°C$ (above this temperature metallation of THF begins). Experimentally it is more convenient, however, to add the solution of BuLi in hexane (dropwise) to a strongly cooled solution of the substrate and *t*-BuOK, in THF or to add the solution of *t*-BuOK to a strongly cooled mixture of the substrate, BuLi, THF, and hexane. Using this combination of basic reagents, a rapid metallation of allylic ethers at temperatures below $-70°C$ is possible. The metallation of allyl *t*-butyl ether is carried out in the procedure below. Alkylations of alkali-metal derivatives of this ether with primary alkyl halides show a γ-regioselectivity that is considerably higher than that observed with other metallated allylic ethers. With *sec*-alkyl halides, however, there is hardly any preference for one of the sites in the metal derivative: *c*-hexyl bromide, for example, gives almost equal amounts of α- and γ-alkylated product. Me_3SiCl and CH_3SSCH_3 show a strong (γ/α ratio > 9) preference for terminal attack. As in the case of the allylic amines, the γ-products are almost exclusively the Z-isomers, a reflection of the thermodynamically preferred configuration of the ether anions [4]. The counterion (Li^+, Na^+, K^+, Cs^+) seems to have little influence [5].

Procedure

A mixture of THF (100 ml), allyl t-butyl ether (0.12 mol, 13.7 g, see Exp. 9), and t-BuOK (0.1 mol, 11.4 g) is placed in the flask. The almost clear solution is cooled to $-80\,^\circ C$ (cooling in a bath with liquid N_2), after which a solution of 0.1 mol of butyllithium in 67 ml of hexane is added over 10 min, while maintaining the temperature between -70 and $-80\,^\circ C$ (occasional cooling). The resulting orange solution is stirred for an additional 10 min at $\sim -75\,^\circ C$, then for 10 min at $-60\,^\circ C$. n-Hexyl bromide (0.1 mol, 16.5 g) is then added over 10 min with vigorous stirring and cooling between -60 and $-50\,^\circ C$. After the addition, the cooling bath is removed and the temperature allowed to rise to $0\,^\circ C$. Water (100 ml) is then added to the almost colourless suspension with vigorous stirring. The product ($\sim 90\%$ Z-γ, 10% α), b.p. $102\,^\circ C/12$ mmHg, $n_D(20)$ 1.4321, is isolated in ~ 85 yield in the manner described for $Me_3SiCH_2CH{=}CH{-}N(C_2H_5)_2$ (see Exp. 1).

Literature

1. Evans DA, Andrews GC, Buckwalter B (1974) J Am Chem Soc 96:5560
2. Clark Still W, Macdonald TL (1974) ibid., 5561
3. Felkin H, Tambuté A (1969) Tetrahedron Lett 821
4. Kloosterziel H, van Drunen JAA (1970) Recl Trav Chim Pays-Bas 89:32
5. Andringa H, Brandsma L (unpublished)

3.3 Lithiation of Allyl Trimethylsilane with BuLi · TMEDA

$$Me_3SiCH_2CH{=}CH_2 + BuLi \cdot TMEDA \xrightarrow[20 \to 40\,^\circ C]{hexane}$$

$$Me_3SiCH(Li)CH{=}CH_2 + C_4H_{10}$$

Functionalization with trimethylchlorosilane and cyclohexanone.

Apparatus: p. 24; Fig. 1; 1 500 ml

Scale: 0.1 molar

Introduction

In order to achieve metallation of allyl trimethylsilane, Schaumann, Chan, and co-workers [1,2] use the strongly basic systems sec-BuLi·TMEDA-THF, n-

BuLi·TMEDA or HMPT–THF and t-BuLi·TMEDA or HMPT–THF. Magnus et al. reported [3] that complete lithiation is effected by BuLi (presumably n-BuLi) in THF at $-40\,°C$ (no reaction time is mentioned). We found that under these conditions no conversion takes place, but observed a very smooth reaction when using the BuLi-TMEDA combination in hexane at 20 to $30\,°C$.

This experiment describes the lithiation of allyl trimethylsilane and the conversions of the lithio compound with trimethylchlorosilane and cyclohexanone, reactions that proceed with very high γ-selectivity. For functionalizations with other electrophiles, e.g., epoxides, alkyl halides, and carbon dioxide, see Ref. [1] and literature cited in this paper.

Whereas in the cases of metallated allylic amines and ethers the products resulting from γ-attack are predominantly Z-isomers, the γ-products obtained from reactions with metallated allyl trimethylsilane are exclusively E-isomers.

Procedure

A solution of 0.1 mol of butyllithium in 67 ml of hexane is placed in the flask. TMEDA (0.10 mol, 11.6 g) is added in one portion at room temperature. The resulting solution is cooled to $20\,°C$ and allyl trimethylsilane (0.11 mol, 12.5 g, see Exp. 11) is added in one portion. The temperature may rise over a few min to $30\,°C$ or higher. The orange-brown solution is kept for 30 min at $40\,°C$, after which functionalization reactions are carried out.

Reaction with trimethylchlorosilane

Et_2O (50 ml) is added and the solution is cooled to $-40\,°C$. Freshly distilled trimethylchlorosilane (0.11 mol, 11.9 g) is added in one portion with vigorous stirring. After 10 min water (100 ml) is added to the resulting white suspension. The organic layer is washed six times with 30-ml portions of water to remove the TMEDA. The combined aqueous layers are extracted once with 50 ml of pentane and the extract is washed twice with water. After drying the combined organic solutions over $MgSO_4$, the solvent is removed under reduced pressure (bath temperature $\leqslant 35\,°C$) and the remaining liquid carefully distilled through a 30-cm Vigreux column. E-$Me_3SiCH_2CH{=}CH$—$SiMe_3$, b.p. $52\,°C/12\,mmHg$, $n_D(20)$ 1.4303, is obtained in greater than 80% yield.

Reaction with cyclohexanone

THF (50 ml) is added and the solution is cooled to $-70\,°C$. Cyclohexanone (0.1 mol, 9.8 g) is added over 5 min with cooling between -70 and $-50\,°C$. After an additional 2 min, water (100 ml) is added with vigorous stirring. The organic layer and two ethereal extracts are dried over $MgSO_4$ and subsequently concentrated under reduced pressure. Distillation of the remaining liquid through a 20-cm Vigreux column gives the carbinol E-$(CH_2)_5(OH)CH_2CH{=}CH$—$SiMe_3$, b.p. $120\,°C/15$ mmHg, $n_D(20)$ 1.4773, in $\sim 65\%$ yield.

Literature

1. Schaumann E, Kirschning A (1988) Tetrahedron Lett 29:4281
2. Lau PWK, Chan TH (1978) Tetrahedron Lett 2383
3. Ayalon-Chass D, Ehlinger E, Magnus P (1977) J Chem Soc Chem Comm 772

3.4 Lithiation of Methyl Allyl Sulfide and Phenyl Allyl Sulfide with BuLi

$$H_2C{=}CHCH_2SR + BuLi \xrightarrow[-80\to-20\,°C]{THF\text{-hexane}}$$
$$H_2C{=}CHCH(Li)SR + C_4H_{10} \quad (R{=}CH_3, Ph)$$

Functionalization with oxirane.

Apparatus: p. 24; Fig. 1; 500 ml

Scale: 0.1 molar

Introduction

Allylic sulfides are metallated much more easily than the analogous amines, ethers, and silanes. Addition of allyl methyl sulfide or allyl phenyl sulfide to a suspension of sodamide in liquid ammonia gives an orange to red solution which reacts vigorously with alkyl halides to give -in excellent yields- a mixture of γ- and α-alkyl derivative, the latter being the main component. pK values of allylic sulfides are not known, but we presume that they are close to the value of ammonia (~ 34). Aryl—or heteroaryl—and vinyl allyl sulfides are probably somewhat more acidic than alkyl allyl sulfides. Interaction between alkali amides and the analogous allylic amines and ethers at best give low concentrations of short-living allylic anions, which are reprotonated to 1-propenylamines and ethers [1]. Although LDA is probably sufficiently strong to effect complete metallation of aryl and alkyl allyl sulfides, butyllithium is preferred since it reacts faster.

Reacions of metallated allylic sulfides with electrophiles that show a high site selectivity (either α or γ) are rare. One example is the β-hydroxyalkylation of lithiated phenyl allyl sulfide with oxirane, giving almost exclusively the α-derivative $H_2C{=}CHCH(SPh)CH_2CH_2OH$ [2].

Procedure

A solution of 0.105 mol of butyllithium in 71 ml of hexane is placed in the flask. THF (80 ml) is added with cooling below 0 °C, after which the solution is cooled to -80 °C. Allyl methyl sulfide (0.11 mol, 9.7 g) or allyl phenyl sulfide (0.1 mol, 15.0 g, see Exp. 10) is added dropwise over 10 min. After an additional 10 min (at -80 °C), the cooling bath is removed and the temperature allowed to rise to $-20\,°C$. The dark-yellow or orange solution is cooled to below $-50\,°C$ after which the electrophile is added. Alkyl bromides and trimethylchlorosilane react almost instantaneously at these low temperatures.

A mixture of 0.15 mol (6.6 g) of oxirane and 30 ml of THF is added over a few s at $-70\,°C$. After removal of the cooling bath, the temperature rises to above $0\,°C$ over a few s and the solution becomes faintly yellow or colourless. Water (100 ml) is then added with vigorous stirring. The organic layer and one ethereal extract are dried over $MgSO_4$ and subsequently concentrated in vacuo. Distillation of the remaining liquid through a very short Vigreux column (5 cm) gives the alcohol $H_2C=CHCH(SPh)CH_2CH_2OH$, b.p. $\sim 155\,°C/3$ mmHg, $n_D(20)$ 1.5758, in greater than 80% yield.

Literature

1. Brandsma L (unpublished observations)
2. Voyle M, Kyler KS, Arseniyadis S, Dunlap NK, Watt DS (1983) J Org Chem 48:470

3.5 Metallation of Methyl Isopropenyl Ether with BuLi · t-BuOK

$$H_2C=C(OCH_3)CH_3 + BuLi \cdot t\text{-}BuOK \xrightarrow[-80 \to -50°C]{\text{THF-hexane}}$$

$$H_2C=C(OCH_3)CH_2K + t\text{-}BuOLi$$

$$H_2C=C(OCH_3)CH_2K + C_8H_{17}Br \longrightarrow$$
$$H_2C=C(OCH_3)CH_2C_8H_{17} + KBr\downarrow$$

Apparatus: p. 24, Fig. 1; 11

Scale: 0.1 molar

Introduction

Treatment of the allylic ethers $CH_3CH=CHCH_2OR$ and $H_2C=C(CH_3)CH_2OR$ with *sec*-BuLi in THF gives only the allylic anions $CH_3CH=CH—CH—OR$ and $H_2C=C(CH_3)—CH—OR$, indicating that OR groups enhance the kinetic acidity of adjacent allylic protons [1, compare, however, 2]. In methyl isopropenyl ether such activated allylic protons are absent and, as expected from the data in the literature, this compound is metallated less easily than allylic ethers $H_2C=CHCH_2OR$ (see Table 6). The conditions required for deprotonation of the isopropenyl ether are comparable with those applied in the metallation of propene and isobutene (see Chap. II). Successful metallation can be achieved only with BuLi · t-BuOK in THF-hexane mixtures at low temperatures using a large excess of the isopropenyl ether [3]. BuLi in THF and BuLi · TMEDA in hexane react sluggishly below $0\,°C$ to give suspensions that are unreactive towards alkyl bromides. Presumably the metallated ether, $H_2C=C(OCH_3)CH_2M$, undergoes some decomposition reaction (which remains to be investigated). For this reason, reactions of the metallated isopropenyl ether have to be carried out at temperatures as low as possible. The metallated derivative obtained by reaction at low temperatures with $BuLi \cdot t\text{-}BuOK$ represents a novel anion-equivalent of the acetone-enolate, $CH_3COCH_2^- \leftrightarrow CH_3C(O^-)=CH_2$. This may be illustrated by the acid hydrolysis of the alkylation products $H_2C=C(CH_2R)OCH_3$, affording methyl ketones, CH_3COCH_2R, in excellent yields.

Procedure

A solution of 0.1 mol of butyllithium in 67 ml of hexane is placed in the flask. THF (40 ml) is added with cooling below 0 °C, after which the solution is cooled to − 90 °C (occasional cooling in a bath with liquid N_2). Subsequently, a solution of 0.1 mol (11.2 g) of t-BuOK in 70 ml of THF is added dropwise over 10 min, while maintaining the temperature in the flask between − 95 and − 85 °C. Freshly distilled methyl isopropenyl ether (0.4 mol, 29 g) is then added at ∼ − 85 °C over a few min. A thick faintly yellow suspension is formed soon. The mixture is stirred (at a moderate rate, splashing being avoided) for 30 min at ∼ − 75 °C, then for the same period at − 50 °C. The suspension becomes much thinner. Octyl bromide (0.09 mol, 17.4 g) is then added over 5 min with cooling between − 50 and − 60 °C. The heating effect is rather strong. After stirring for an additional 10 min at − 50 °C, the cooling bath is removed and the temperature allowed to rise above 0 °C. Water (150 ml) is added with vigorous stirring. The organic layer and two ethereal extracts are dried over $MgSO_4$ and subsequently concentrated under reduced pressure. Careful distillation of the remaining liquid through a 30-cm Vigreux column gives the alkylation product, $H_2C=C(OCH_3)CH_2C_8H_{17}$, b.p. 100 °C/12 mmHg, $n_D(20)$ 1.4370, in ∼ 85% yield.

Literature

1. Evans DA, Andrews GC, Buckwalter B (1974) J Am Chem Soc 96:5560
2. Hunter DA, Lin Y, McIntyre AL, Shearing DJ, Zvagulis M (1973) J Am Chem Soc 95:8327
3. Heus-Kloos YA, Brandsma L (unpublished)

3.6 α-Metallation of Benzyl Dimethylamine with BuLi · t-BuOK

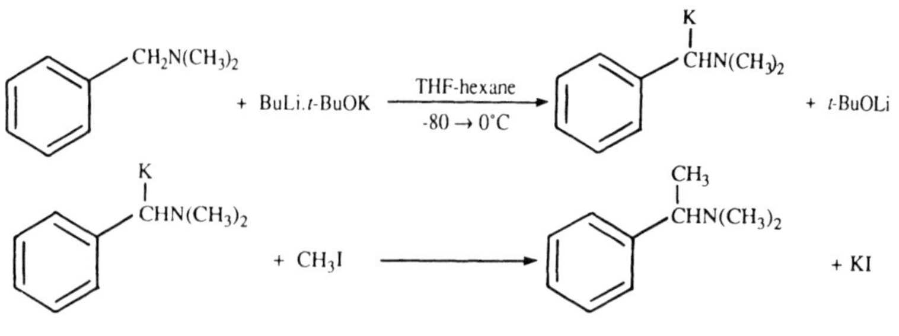

Apparatus: p. 24; Fig. 1; 11

Scale: 0.1 molar

Introduction

Hauser et al. [1] investigated the metallation of N,N-dimethyl benzylamine with a number of strong bases. Whereas BuLi and BuNa gave either exclusively the O-metallated or α-metallated derivative, the reaction with PhLi, PhNa, and PhK led directly to the α-metallated product. A similar result is obtained when treating the amine with BuLi · t-BuOK ($\equiv$'BuK') in a mixture of THF and hexane [2]. The dark-red solution thus obtained reacts very smoothly with alkyl halides even at very low temperatures. In fact, the reaction with methyl iodide may be carried out like a titration using the characteristic colour as indicator.

Procedure

To a mixture of 0.1 mol (13.5 g) of N,N-dimethyl benzylamine and 40 ml of THF is added a solution of 0.11 mol of butyllithium in 74 ml of hexane with cooling below $-50\,°C$. A solution of 0.11 mol (12.5 g) of t-BuOK in 60 ml of THF is added over 15 min while keeping the temperature of the dark-red solution between -70 and $-80\,°C$ (occasional cooling in a bath with liquid N_2). After an additional 30 min, the cooling bath is removed and the temperature allowed to rise to $0\,°C$, then the solution is cooled to $\sim -80\,°C$ and about 0.10 mol (14.2 g) of methyl iodide is added dropwise over about 10 min while maintaining the temperature between -80 and $-70\,°C$. At the last stage of the addition, the colour disappears. The white suspension is hydrolyzed with 100 ml of water. The organic layer and three ethereal extracts are dried over potassium carbonate and then concentrated in vacuo. Careful distillation of the remaining liquid through a 30-cm Vigreux column gives the methylation product, b.p. $70\,°C/12\,mmHg$, $n_D(20)$ 1.5030, in $\sim 65\%$ yield.

Literature

1. Puterbaugh WH, Hauser CR (1963) J Am Chem Soc 85: 2467
2. Heus-Kloos YA, Brandsma L (unpublished)

3.7 Lithiation of Methyl Benzyl Sulfide and Benzyl Trimethylsilane

$$PhCH_2SCH_3 + BuLi \xrightarrow[-50 \to -20°C]{\text{THF-hexane}} PhCH(Li)SCH_3 + C_4H_{10}$$

$$PhCH(Li)SCH_3 + Me_3SiCl \longrightarrow PhCH(SiMe_3)SCH_3 + LiCl\downarrow$$

$$PhCH_2SiMe_3 + BuLi \cdot TMEDA \xrightarrow[20 \to 45°C]{\text{hexane}} PhCH(Li)SiMe_3 + C_4H_{10}$$

$$PhCH(Li)SiMe_3 + Me_3SiCl \longrightarrow PhCH(SiMe_3)_2 + LiCl\downarrow$$

$$PhCH(Li)SiMe_3 + cyclohexanone \longrightarrow$$

$$(CH_2)_5C(OLi)-CH(Ph)SiMe_3 \xrightarrow{-Me_3SiOLi} (CH_2)_5C{=}CHPh$$

Apparatus: p. 24, Fig. 1; 1 500 ml

Scale: 0.1 molar

Introduction

In this experiment the conditions for the α-lithiation of methyl benzyl sulfide and benzyl trimethylsilane are compared. Whereas the sulfide is readily metallated at very low temperature by BuLi in a THF-hexane mixture, the silyl derivative does not react at all, and even at $\sim 0-20\,°C$ the metallation proceeds sluggishly. Using the more efficient couple BuLi·TMEDA in hexane, however, a rapid lithiation of the silyl compound can be effected at room temperature: under these conditions toluene reacts much more slowly. Thus, a trimethysilyl group activates protons in adjacent positions, but to a much lesser extent than does a methylthio group.

The intermediates obtained react in the usual way with various electrophiles to give the expected derivatives in excellent yields. The hydroxyalkylation of lithiated benzyl trimethylsilane is immediately followed by a Peterson-elimination of Me_3SiOLi to give the olefin.

Procedure

Metallation of methyl benzyl sulfide and subsequent trimethylsilylation

A solution of 0.105 mol of BuLi in 71 ml of hexane is placed in the flask. THF (80 ml) is added with cooling below $- 50\,°C$. Subsequently, methyl benzyl sulfide (13.8 g, 0.1 mol, see Exp. 10) is added over 5 min with cooling between $- 50$ and $- 40\,°C$. The cooling bath is removed and stirring at $- 20\,°C$ is continued for 10 min, then trimethylchlorosilane (0.11 mol, 11.9 g, freshly distilled) is added to the brown solution at $\sim - 20\,°C$ over a few min. The white suspension (sometimes colourless solution) is then hydrolyzed and the aqueous layer extracted with Et_2O. After drying the solution over $MgSO_4$, the solvent is removed in vacuo and the remaining liquid distilled through a short Vigreux column to give the expected product, b.p. 150 °C/12 mmHg, $n_D(20)$ 1.5758, in an excellent yield.

Metallation of benzyl trimethylsilane and subsequent reaction with trimethylchlorosilane and cyclohexanone

To a solution of 0.105 mol of BuLi in 71 ml of hexane is added at room temperature 0.11 mol (12.8 g) of TMEDA. The solution is cooled to 20 °C and 0.1 mol (16.4 g) of benzyl trimethylsilane (Chap. II, Exp. 1) is added in one portion. The temperature rises within a few min to above 35 °C. After an additional 10 min (at 45 °C), the red solution is cooled to below 0 °C (a yellowish suspension is formed). Et_2O (50 ml) is added, after which freshly distilled trimethylchlorosilane (0.11 mol, 11.9 g) is added over a few min with vigorous stirring and cooling below 0 °C. The white suspension is then hydrolyzed with water.

For the reaction with cyclohexanone, the solution of the lithiated silane in Et_2O-hexane is cooled to -60 °C, after which cyclohexanone (0.11 mol, 10.8 g) is added in one portion with vigorous stirring. The temperature may rise to at least -10 °C over a few s. After stirring the solution for 15 min at room temperature, the reaction mixture is hydrolyzed and the aqueous layer extracted once with Et_2O.

In both cases the organic solution is dried over $MgSO_4$ and subsequently concentrated in vacuo. Distillation through a 20-cm Vigreux column gives $PhCH(SiMe_3)_2$, b.p. 108 °C/12 mmHg, $n_D(20)$ 1.5006, and $(CH_2)_5C{=}CHPh$, b.p. 125 °C/12 mmHg, $n_D(20)$ 1.5585, in greater than 80% yields.

3.8 Preparation of *N,N*-Dimethyl Allylamine and *N,N*-Diethyl Allylamine

$$H_2C{=}CHCH_2Br + 2Me_2NH \xrightarrow[5\,^-40\,°C]{\text{Petroleum ether}}$$

$$H_2C{=}CHCH_2NMe_2 + Me_2NH \cdot HBr\downarrow$$

$$H_2C{=}CHCH_2Br + 2Et_2NH \xrightarrow[\text{reflux}]{E_2O}$$

$$H_2C{=}CHCH_2NEt_2 + Et_2NH \cdot HBr\downarrow$$

Apparatus: for the reaction with Me_2NH: p. 24, Fig. 1, 2l; for the reaction with Et_2NH: 1-l, three-necked, round-bottomed flask, equipped with a dropping funnel, a mechanical stirrer and a reflux condenser.

Scale: 1.0 molar

Procedure

$H_2C{=}CHCH_2NMe_2$

High-boiling petroleum ether (500 ml, b.p. > 170 °C/760 mmHg) is cooled to 0 °C, after which 2.1 mol (95 g) of liquified dimethylamine is added. Allyl bromide (1.0 mol, 121 g) is added dropwise over 30 min with efficient stirring and cooling between 5 and 10 °C. A thick white suspension is formed gradually. After the addition, the cooling bath is removed and the temperature allowed to rise (occasional cooling may be necessary if the temperature rises too fast). The mixture is stirred for an additional 1 h at 40 °C, then the equipment is removed, two stoppers are placed on the outer necks and the flask is equipped for a vacuum distillation: 40-cm Vigreux column, condenser, and single receiver cooled in a bath with dry ice and acetone (<-70 °C). A tube filled with KOH pellets is placed between the apparatus and the water aspirator. The system is evacuated (10–20 mmHg) and the flask gradually heated. The evacuation procedure is stopped when the petroleum ether begins to distill (b.p. > 50 °C/10–20 mmHg). Careful distillation of the contents of the receiver at atmospheric pressure through a 40-cm Vigreux column gives the allylic amine, b.p. 60 °C/760 mmHg, $n_D(20)$ 1.4001, in at least 70% yield.

$H_2C{=}CHCH_2NEt_2$

Allyl bromide (1.0 mol, 121 g) is added portionwise over 25 min to a vigorously stirred mixture of 300 ml of Et_2O and 2.2 mol (161 g) of diethylamine (dried over a large amount of KOH pellets for one night or over machine-powdered KOH for a few min). After a short period the ether begins to reflux. Refluxing is continued (external heating) for an additional 3 h, then the suspension is cooled to room temperature and ice water ($\sim$ 120 ml) is added with vigorous stirring. The upper layer and one ethereal extract are dried over potassium carbonate (50 g). After filtration and rinsing of the drying agent with Et_2O, the Et_2O is slowly distilled off through an efficient column. The allylic amine, b.p. 113 °C/760 mmHg, $n_D(20)$ 1.4189, is obtained in $\sim$ 75% yield.

3.9 Prepration of *t*-Butyl Allyl Ether

$$H_2C{=}CHCH_2Br + t\text{-}BuOK \xrightarrow[<-33\,°C]{\text{liq. } NH_3}$$

$$H_2C{=}CHCH_2Ot\text{-}Bu + KBr\downarrow$$

Apparatus: p. 24, Fig. 1; 11

Scale: 1.0 molar

Procedure

Anhydrous liquid ammonia of good quality (500 ml, water content less than 0.1%) is placed in the flask. Potassium *tert*-butoxide (1.4 mol, 157 g) is cautiously added with gentle stirring through a powder funnel. The solution is cooled to $\sim -40\,^\circ\text{C}$ (dry ice/acetone) while a slow stream of nitrogen is passed through the apparatus. Allyl bromide (1.0 mol, 121 g) is added in 10-g portions over 30 min while maintaining the temperature between -35 and $-40\,^\circ\text{C}$. After the addition the cooling bath is removed and stirring is continued for 1 h. High-boiling petroleum ether (300 ml, b.p. $> 190\,^\circ\text{C}$ at atmospheric pressure) is then added over a few min with vigorous stirring. The reaction mixture is cautiously poured onto 1 kg of finely crushed ice in a wide-necked conical or round-bottomed flask. After separation of the layers, the aqueous layer is extracted twice with 100-ml portions of petroleum ether. The combined organic solutions are washed five times with 200-ml portions of water and subsequently four times with 100-ml portions of 2 N hydrochloric acid in order to remove t-butylalcohol. The t-butyl ether, b.p. $102\,^\circ\text{C}/760\,\text{mmHg}$, $n_D(20)$ 1.4018, is isolated in 60–64% yield after vacuum distillation (receiver cooled at $-78\,^\circ\text{C}$) and subsequent distillation at atmospheric pressure.

3.10 Methyl Allyl Sulfide, Methyl Benzyl Sulfide and Phenyl Allyl Sulfide

$$CH_3SK + H_2C{=}CHCH_2Br \xrightarrow{\text{CH}_3\text{OH}} CH_3SCH_2CH{=}CH_2 + KBr\downarrow$$

$$PhSK + H_2C{=}CHCH_2Br \xrightarrow{\text{CH}_3\text{OH}} PhSCH_2CH{=}CH_2 + KBr\downarrow$$

$$CH_3SK + PhCH_2Br \xrightarrow{\text{CH}_3\text{OH}} CH_3SCH_2Ph + KBr\downarrow$$

Apparatus: p. 24, Fig. 1; 11

Scale: 0.4 molar

Procedure

A solution of technical potassium hydroxide (corresponding to 0.5 mol) in 250 ml of methanol is cooled to $-20\,^\circ\text{C}$, after which the thiol (0.5 mol, dissolved in 50 ml of cold ($0\,^\circ\text{C}$) methanol) is added over a few min, immediately followed (over 10 min) by allyl bromide (0.4 mol, 48.4 g) or benzyl bromide (0.4 mol, 69.0 g). During these additions the temperature is kept between 0 and $20\,^\circ\text{C}$. After an additional 10 min, 500 ml of an aqueous solution of 20 g of potassium hydroxide is added. The mixture is extracted five times with small portions of pentane (in the case of methyl allyl sulfide) or a 1:1 Et_2O-pentane mixture (in the other cases). The extracts are washed with water and then dried over $MgSO_4$. Methyl allyl sulfide (b.p. $98\,^\circ\text{C}/760\,\text{mmHg}$, $n_D(22)$ 1.4702) is isolated by distillation through a 40-cm Vigreux column, the other sulfides are isolated by vacuum distillation: b.p. $PhSCH_2CH{=}CH_2$ $97\,^\circ\text{C}/15\,\text{mmHg}$, $n_D(22)$ 1.5744; b.p. $PhCH_2SCH_3$ $82\,^\circ\text{C}/12\,\text{mmHg}$, $n_D(22)$ 1.5631.

3.11 Preparation of Allyl Trimethylsilane

$$H_2C=CHCH_2Br + Mg \xrightarrow[0-5\,°C]{Et_2O} H_2C=CHCH_2MgBr$$

$$H_2C=CHCH_2MgBr + Me_3SiCl \longrightarrow H_2C=CHCH_2SiMe_3 + MgBrCl$$

Apparatus: p. 24, Fig. 1; 11

Scale: 1.0 molar (Me₃SiCl)

Procedure

Magnesium turnings (48 g, 2 mol) and Et₂O (600 ml) are placed in the flask. Mercury(II) chloride (2 g) is added and the mixture is stirred for 30 min at room temperature, then the flask is placed in a bath with crushed ice and ice water. As soon as the temperature in the flask has become 0 °C, 5 ml of the 1.3 mol (157 g) of allyl bromide is added. Usually the reaction starts within a few min as is visible from a distinct rise of the temperature in the flask to between 5 and 10 °C. When the exothermic reaction has subsided, the remainder of the allyl bromide is added over 1 h, while keeping the temperature between 2 and 7 °C. After an additional 30 min, the resulting black-grey solution of allylmagnesium bromide is transferred (preferably by canula under a light pressure) into another 1-l flask (filled with inert gas) and the excess of magnesium is rinsed three times with 70-ml portions of Et₂O. Freshly distilled trimethylchlorosilane (1.0 mol, 109 g) is then added dropwise over 30 min. Initially, the temperature is kept between − 15 and − 10 °C, but in the course of the addition it is allowed to rise gradually to 15–20 °C. The conversion is terminated by heating under reflux for 30 min. The reaction mixture is cooled to room temperature and subsequently poured into 500 ml of a cold aqueous solution of 100 g of ammonium chloride. After successive swirling and shaking, the layers are separated and the aqueous layer is extracted once with 50 ml of pentane. After drying over MgSO₄, the greater part of the solvent is distilled off very slowly (over 1 day) through a 40- to 50-cm Widmer column. The remaining liquid is distilled through a shorter (30-cm) Widmer column to give allyl trimethylsilane, b.p. 84 °C/760 mmHg, $n_D(20)$ 1.4068, in yields between 60 and 70%.

Chapter VI
Metallation of Heterocyclic Compounds

1 Introduction

This chapter deals with the metallation of 2-methylthiazoles, -thiazolines, -dihydrooxazines and a number of methylpyridines. In all cases lateral metallation is more easy than removal of an sp^2-proton. In this respect, the behaviour of 2-methylthiazoles and of methylpyridines is completely different from that of methylthiophenes and -furans, which are metallated exclusively on an sp^2-carbon atom. The presence of the aza-nitrogen atom in the aromatic rings is also responsible for an increased acidity of the methyl protons compared to those in methylbenzenes, particularly when the CH_3 group is in the '*ortho*'- or '*para*'-position of nitrogen. This can be ascribed to the possibility of charge-delocalization in the anions, the more electronegative nitrogen atom giving rise to a considerable stabilization.

The use of metallated thiazoles, thiazolines, oxazolines, and dihydrooxazines as anion-equivalents in organic syntheses has been advanced extensively [1–6]. Table 7 summarizes the conditions for 'preparative' metallations of a numbers of substituted heterocycles.

Literature

1. Altman LJ, Richheimer SL (1971) Tetrahedron Lett. 4709; compare Knaus G, Meyers AI (1974) J Org Chem 39:1192
2. Corey EJ, Boger DL (1978) Tetrahedron Lett 5
3. Meyers AI, Durandetta JL, Munavu R (1975) J Org Chem 40:2025
4. Meyers AI, Nabeya A, Adickes HW, Politzer IR, Malone GR, Kovelesky AC, Nolen RL, Portnoy RC (1973) J Org Chem 38:36
5. Schmidt RR (1972) Synthesis 333
6. Meyers AI, Mihelich ED (1976) Angew Chem 88:321; Angew Chem, Int ed 15:499

2 Metallation of Alkyl Derivatives of Pyridine and Quinoline

The pK values of the three methylpyridines determined by Fraser et al. [1] show marked differences: 2- and 4-methylpyridine (pK $\sim$ 34 and 32, respectively) are much more acidic than 3-picoline (pK $\sim$ 37) because the negative charge in the anions can reside on nitrogen, which is not possible in the 3-picolyl anion [2]. These differences are reflected in the behaviour of the compounds towards strongly basic reagents. LDA effects complete lithiation of 2- and 4-methylpyridine as may be concluded from the excellent yields in derivatizations with a variety of electrophiles, including alkyl halides, carbonyl compounds, trimethylchlorosilane, and dialkyl disulfides (see exp. section). In the case of 3-picoline, good yields are obtained only with alkyl halides. Other electrophiles give lower, and in some cases poor yields [3, 4]. LDA as well as alkali amides (in liquid ammonia) are not sufficiently strong to effect complete metallation of 3-picoline and certain electrophiles will react at comparable rates with the amide and the picolyl anion, occurring in the equilibrium

(R= H or alkyl)
(M= Li, Na, K)

The good yields in the reaction of alkyl bromides with 3-picolyl-lithium (from 3-picoline and LDA in THF) may be explained by assuming that the alkyl halides react faster with 3-picolyl lithium (a super-nucleophile, see Chap. I, p. 2) than with LDA. Analogous situations arise when toluene, isoprene and α-methylstyrene are treated with potassium diisopropylamide and an alkyl halide is added (see Chap. II). 2-Picoline can be readily metallated with a variety of base-solvent combinations including LDA–Et$_2$O, LDA–THF, BuLi–Et$_2$O or THF, NaNH$_2$ or KNH$_2$-liquid ammonia (but not with LiNH$_2$-liquid ammonia with which the ionization equilibrium is strongly on the side of the reactants). For 4-picoline, LDA and all alkali amides are suitable. With BuLi (in Et$_2$O or THF) the metallation of this compound does not proceed cleanly, addition of the base across the azomethine double bond being a serious side-reaction [4].

Introduction of a second or third methyl group into the pyridine ring is expected to give rise to a slight increase of the pK (presumably not more than 0.5 to 1 unit). Nevertheless LDA, sodamide, and potassium amide will cause complete or extensive 'ionization'. The behaviour of the various isomeric dimethyl- and trimethyl-pyridines can be predicted in part from experimental data on the metallation of the picolines.

a. CH$_3$-groups at the 3- and 5-positions will remain unattacked.
b. CH$_3$-groups at the 4-position are more acidic in a thermodynamic sense than CH$_3$-groups at the 2- and 6-positions. Thus, under equilibrium conditions 2,4-lutidine will be completely converted to the 4-metallated product when using *one* molar equivalent of base. In the presence of a proton-donor (this may be ammonia, diisopropylamine or even 2,4-lutidine itself) the 2-lutidyl anion may isomerize to the 4-lutidyl anion by a process of proton-donation and-removal [2]. Similar processes may be expected or have been reported [2, 5] for the metallation of the various collidines (trimethylpyridines) and dialkylquinolines. In THF and Et$_2$O the kinetically favoured metallation (by LDA or BuLi) of 2, 4-lutidine is on the 2-methyl group. This may be explained by assuming the formation of an initial coordination complex of the base [2]

In strongly polar solvents, e.g., in liquid ammonia or with HMPT present as a co-solvent such a chelate is not stable and the 4-lutidyl anion may be formed directly.

2,6-Lutidine has been 2,6-dimetallated with two mol equivalents of BuLi·t-BuOK [6, 7]:

The dimetallation of 2,4-lutidine can be achieved under similar conditions but proceeds less cleanly [4].

Literature

1. Fraser RR, Mansour TS, Savard S (1985) J Org Chem 50:3232
2. Kaiser EM (1983) Tetrahedron 2055
3. Kaiser EM, Petty JD (1975) Synthesis 705
4. Brandsma L, Verkruijsse HD (unpublished observations)
5. Kaiser EM, Bartling GJ, Thomas WR, Nichols SB, Nash DR (1973) J Org Chem 38:71
6. Bates RB, Ogle CA (1982) J Org Chem 48:3949
7. Hacker R, von R Schleyer P, Reber G, Müller G, Brandsma L (198) J Organometal. Chem 316:C4

3 Lateral Metallation of 2-Substituted Oxazolines, Thiazolines, Dihydrooxazines, and Thiazoles

Although the base-strength of LDA and alkali amides (in liquid ammonia) is sufficient for complete metallation of the 2-substituted heterocycles, butyllithium is generally used. With this reagent metallation is readily achieved at very low temperatures. Especially in the case of 2-alkyldihydrooxazines this is an essential condition in connection with the low stability of the 2-metallated derivatives (irreversible ring opening, see Sect. 1 for references):

There are no indications for an analogous ring opening of 2-lithiomethylthiazoline at temperatures below $-10\,°C$.

Table 7. Representative conditions for the lateral metallation of heterocyclic compounds*

Substrate	Base–solvent	Temp. range** (°C)	Notes
2-CH$_3$-pyridine	Na(K)NH$_2$–liq. NH$_3$	$-33\,(0)$	a
	LDA–Et$_2$O or THF, hexane	$-30\,(0) \to +20\,(15)$	
	BuLi–Et$_2$O or THF, hexane	$-50\,(0) \to -20\,(0)$	b
3-CH$_3$-pyridine	KNH$_2$–liq. NH$_3$	$-33\,(0)$	c
	LDA–THF–HMPT, hexane	$0\,(30)$	d
4-CH$_3$-pyridine	Na(K)NH$_2$–liq. NH$_3$	$-33\,(0)$	
	LDA–Et$_2$O or THF, hexane	$-20\,(0) \to +20\,(10)$	e
2,6-(CH$_3$)$_2$-pyridine	BuLi–THF or Et$_2$O, hexane	$-50\,(0) \to -20\,(0)$	b
	NaNH$_2$ or KNH$_2$–liq. NH$_3$	-33	
2,6-(CH$_3$)$_2$-pyridine dimetallation	BuLi–t-BuOK · TMEDA, hexane (30 mol% excess)	$-50\,(0) \to +10\,(0)$	
2,4-(CH$_3$)$_2$-pyridine metallation on 2-CH$_3$ group	BuLi–Et$_2$O or THF, hexane	$-50\,(0) \to -20\,(0)$	
2,4-(CH$_3$)$_2$-pyridine metallation on 4-CH$_3$ group	KNH$_2$–liq. NH$_3$	$-33\,(0)$	f
	LDA–THF, hexane	$-20\,(0) \to +20\,(15)$	
2,4,6-(CH$_3$)$_3$-pyridine metallation on 4-CH$_3$ group	KNH$_2$–liq. NH$_3$	$-33\,(0)$	g
	LDA–THF, hexane	$-20\,(0) \to +20\,(0)$	
metallation on 2-CH$_3$ group	BuLi–THF or Et$_2$O, hexane	$-40\,(0) \to -10\,(0)$	b, g
2,4-(CH$_3$)$_2$-quinoline metallation on 2-CH$_3$ group	BuLi, THF or Et$_2$O hexane	$-40\,(0) \to -20\,(0)$	b, g
2,4-(CH$_3$)$_2$-quinoline metallation on 4-CH$_3$ group	NaNH$_2$–liq. NH$_3$	-33	g
	LDA–THF, hexane	$-20\,(0) \to +20\,(0)$	
2,4-(CH$_3$)$_2$-thiazole metallation on 2-CH$_3$ group	BuLi–THF, hexane	$-78\,(25)$ (see Sect. 1, Ref. 1)	b, h
2-CH$_3$-thiazoline	BuLi–THF, hexane	$-70\,(0) \to -50\,(5)$	b, i
2,4,4,6-(CH$_3$)$_4$- 5,6-dihydro-1,3-oxazine	BuLi–THF, hexane	$-75\,(60)$	j

* Initial concentrations of base $\sim 0.8\,\text{mol/l}$ (for reactions in organic solvents) or 1 to 1.4 mol/l (for reactions in liquid NH$_3$). In general 5 to 10% excess of substrate is used. Reaction times and temperatures are taken from our experiments except for those for the metallations of 2,4-dimethylthiazole.

** An indication like $-80\,(20) \to -60\,(10)$ means: after a relatively quick addition of the substrate at $-80\,°\text{C}$ the mixture is kept for 20 min at this temperature, then the cooling bath is removed and the temperature allowed to rise to $-60\,°\text{C}$, at which temperature the mixture is stirred for 10 min before carrying out the functionalization reactions.

Notes

a. With lithium amide the metallation is very incomplete.
b. The base is added to a mixture of the substrate and Et$_2$O or THF.
c. The 'ionization' is very incomplete as indicated by the low yield ($\sim 30\%$) 3-pentylpyridine obtained by addition of butyl bromide.
d. Also in this case, the metallation is probably incomplete and successful derivatization reactions are confined to alkylations, condensations with non-enolizable-aldehydes and ketones and (presumably) epoxides (see Sect. 2, Ref. [3]).

4 Experiments

All temperatures are internal, unless indicated otherwise.
All reactions are carried out in an atmosphere of inert gas, except those in boiling ammonia.
For general instructions concerning handling organolithium reagents, drying solvents, etc. see Vol. 1.

4.1 General Procedure for the Metallation of Mono-, Di-, and Trimethylpyridines and Quinolines in Liquid Ammonia and the Subsequent Alkylation

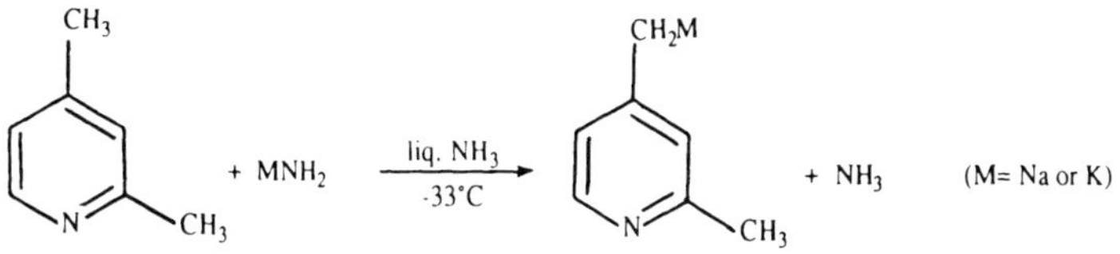

Analogous for other methylpyridines and -quinolines.

Apparatus: 1-l round-bottomed, three-necked flask, equipped with a dropping funnel, a mechanical stirrer and a vent (e.g., rubber stopper with a hole of at least 5 mm diameter).

Scale: 0.3 molar

Introduction

The pK values of 2-, 3-, and 4-methylpyridine have been determined to be ca. 34, 37, and 32, respectively (Sect. 2, Ref. [1]). Di- and trimethylpyridines are expected to be somewhat less acidic. Since the pK value of ammonia lies within this range, interaction between the heterocyclic compounds and the alkali amides will give rise to certain equilibrium concentrations of the metallated pyridine derivatives. The extent of ionization is expected to decrease in the following order: $4-CH_3 > 2-CH_3 > 3-CH_3$. As in the metallations of isoprene and α-methylstyrene

---

e. Reaction of 4-picoline with BuLi gives also rise to addition across the C=N and (Brandsma L, unpublished observation).
f. Initially mainly 2-lithiolutidine is formed at low temperatures.
g. Compare Sect. 2, Ref. [5].
h. Compare Sect. 1, Ref. [1].
i. Compare Meyers AI, Durandetta JL (1975) J Org Chem 40:2021
j. The dihydrooxazine was quickly added (cooling in a bath with liquid N_2) at ~ −80 °C, after which the mixture was stirred for one hour at −75 °C; compare Sect. 1, Ref. [4], in which reversed-order addition is applied.

with alkali diisopropylamide (see Chap. II), the position of the metallation equilibrium will strongly depend on the size of the counterion. This phenomenon generally occurs when compounds with comparable acidities are involved in the metallation equilibrium. Reaction of alkylpyridines with potassium amide will give rise to higher concentrations of the pyridine anions than in the cases of lithium- and sodium amide. The influence of the counterion is nicely illustrated with the reactions of 2-picoline with lithium amide and potassium amide or sodamide. If 2-picoline is added to a suspension of $LiNH_2$ in liquid ammonia the suspended material does not dissolved and only a faint pink colour of the picolyl anion becomes visible (in the case of the more acidic 4-picoline most of the suspended $LiNH_2$ dissolves). Subsequent addition of butyl bromide gives 2-pentylpyridine in a moderate yield. With $NaNH_2$ and KNH_2 an orange solution is formed which reacts vigorously with alkyl halides to give 2-alkylpyridines in high yields. In the case of 3-methylpyridine (the least acidic mono-methylpyridine), the concentration of the anion in liquid ammonia is very small, even when potassium amide is used as a base: addition of an alkyl bromide to the resulting ammoniacal solutions gives 3-alkylpyridines in yields varying from a few % for $LiNH_2$ to ca. 40% for KNH_2. The more acidic 4-picoline gives good results with all the three alkali amides. Addition of a dimethyl- or trimethylpyridine to sodamide or potassium amide in liquid ammonia leads to direct formation of the thermodynamically most stable anion. Thus, 2,4-lutidine and $NaNH_2$ or KNH_2 give the 4-anion. It is not known whether this deprotonation is preceded by the formation of a very short-living 2-metallated intermediate: with LDA in THF 2,4-lutidine gives at low temperatures mainly the 2-lithio derivative. Above $-10\,°C$ rapid isomerization into 4-lithio lutidine takes place. Metallated pyridines and quinolines are extremely reactive towards alkylating reagents. Even with cyclohexyl bromide and bromo-acetaldehyde diethylacetal, $BrCH_2CH(OC_2H_5)_2$, compounds which usually undergo dehydrohalogenation in reactions with nucleophilic species, reasonable or good yields of coupling products are obtained.

Procedure

A suspension of 0.35 mol of sodamide or solution of 0.35 mol of potassium amide in 400 ml of liquid ammonia is prepared from 0.35 mol (8.1 g) of clean sodium or 0.35 mol (13.7 g) of clean potassium as described in Vol. 1, Chap. I. Lutidine (0.3 mol, 32.2 g, dried by shaking with machine-powdered KOH and subsequently filtered) is added over 5 min with stirring at a moderate rate. After an additional 5 min, the alkylating agent (0.4 mol, note 1) is added over 15 min to the vigorously stirred red to orange solution. The reactions are very vigorous, with primary alkyl bromides the conversions proceed almost instantaneously. The additional reaction times in the cases of cyclohexyl bromide and bromoacetaldehyde diethylacetal are not longer than 15 min. After completion of the conversions, the ammonia is removed by placing the flask in a water bath at $\sim 35\,°C$ (the vent is removed, stirring is continued until most of the ammonia has evaporated), or is allowed to evaporate (the equipment is removed and plugs of cotton wool are placed on the necks). Water

(200 ml) is added, after which three to five extractions with Et$_2$O are carried out. If *potassium* amide has been used, however, dichloromethane (200 ml) and water (200 ml) are successively added (note 2). The aqueous layer is extracted twice with dichloromethane.

The organic solution is dried over potassium carbonate and subsequently concentrated under reduced pressure. The product is isolated by careful fractional distillation of the remaining liquid. The following compounds have been prepared by this procedure:

2-Methyl-4-butylpyridine, b.p. 95 °C/12 mmHg, n$_D$(20) 1.4906, yield ~ 80%, from 2,4-lutidine, sodamide and propyl bromide;

2-Methyl-6-butylpyridine, b.p. 87 °C/10 mmHg, n$_D$(21) 1.4882, yield 75%, from 2,6-lutidine, potassium amide and propyl bromide;

2-Butylpyridine, b.p. 80 °C/10 mmHg, n$_D$(21) 1.4902, yield 78%, from 2-picoline, potassium amide and propyl bromide;

4-(1,1-Diethoxypropyl)pyridine, b.p. 150 °C/12 mmHg, n$_D$(21) 1.4804, yield 60%, from 4-picoline, sodamide and BrCH$_2$CH(OC$_2$H$_5$)$_2$;

4-(Cyclohexylmethyl)pyridine, b.p. 130 °C/12 mmHg, n$_D$(20) 1.5192, yield 65%, from 4-picoline, sodamide and cyclohexyl bromide.

Notes

1. The volatile ethyl bromide and propyl bromide are used in a larger excess (0.5 and 0.45 mol, respectively).
2. Small particles of potassium covered by a crust of oxide may remain unconverted. When Et$_2$O is used for the extraction they may cause fire hazards.

4.2 Conversion of 2-Methylpyridine, 2,4-Dimethylpyridine, 2,6-Dimethylpyridine, and 2,4,6-Trimethylpyridine into the 2-Lithiomethyl Derivatives

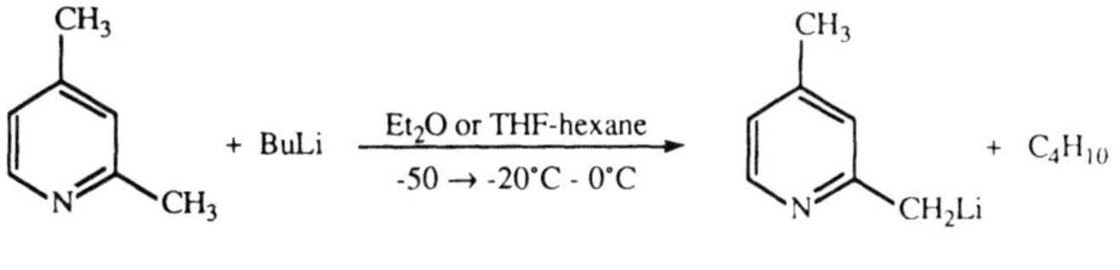

Analogous for other pyridines with a 2-methyl group.
Functionalization with various electrophilic reagents.

Apparatus: p. 24, Fig. 1; 500 ml

Scale: 0.1 molar

Introduction

Although 4-methylprotons in methylpyridines are more acidic in a thermodynamic sense than methyl protons in the 2-position, the proton abstraction from the latter is kinetically favoured. Thus, treatment of a pyridine derivative having methyl groups in the 2- and the 4-position with a strong base initially gives the intermediate with the metal on the 2-methyl group. If the medium has a relatively low polarity (Et$_2$O or THF) and no proton-donating compounds are present (e.g., diisopropylamine, formed when LDA is used as a metallating reagent), the intermediate is reasonably stable and successful regiospecific functionalizations of the 2-methyl group are possible. The most suitable base-solvent combination for the regiospecific generation of 2-methyl-metallated derivatives is butyllithium-THF or Et$_2$O. In the presence of an excess of the methylpyridine, some 4-methyl-lithiated derivative may be formed by a process of proton-donation and -abstraction (the free pyridine derivative being the proton donor). It seems therefore advisable to add the methylpyridine to the butyllithium solution and not to carry out inversed-order addition. The temperature should be kept as low as possible during the metallation and the time interval between metallation and functionalization preferably kept within some fifteen minutes.

Procedure

THF (80 ml) or Et$_2$O (100 ml) is placed in the flask. A solution of 0.105 mol of butyllithium in 70 ml of hexane is introduced by syringe while keeping the temperature of the mixture below 0 °C. After cooling to -50 °C, 0.1 mol of the 2-methylpyridine derivative is added dropwise over 10 min, then the cooling bath is removed and the temperature allowed to rise to -20 °C (0 °C in the case of Et$_2$O). The red solution or suspension can immediately be used for derivatization reactions.

a) Methylthiolation with CH$_3$SSCH$_3$

The cold (-20 °C) solution or suspension is added over 10 min (by syringe) to a vigorously stirred mixture of 0.3 mol (28.2 g) of dimethyldisulfide and 80 ml of THF or Et$_2$O kept between -30 and -60 °C. After an additional 2 min, water (50 ml) is added with vigorous stirring. The aqueous layer is extracted twice with Et$_2$O. After drying the organic solution over potassium carbonate and subsequent concentration under reduced pressure, the product is isolated by distillation through a 30-cm Vigreux column. 2-Methylthiomethylpyridine, b.p. 104 °C/12 mmHg, n$_D$(20) 1.5629, is obtained from 2-methylpyridine in greater than 80% yield. If the addition is carried out in the normal sense, the mono-methylthio derivative is obtained in a low yield, the main product being bis(methylthiomethyl)pyridine (for the explanation see Chap. I, Sect. 9).

b) Trimethylsilylation

Freshly distilled trimethylchlorosilane (0.11 mol, 11.9 g) is added in one portion at -50 °C. The cooling bath is removed and the temperature allowed to rise to 0 °C.

Water (50 ml) is then added to the almost colourless suspension and the product is isolated as described under a. 4-Methyl-2-(trimethylsilylmethyl)pyridine, b.p. 90 °C/12 mmHg, $n_D(20)$ 1.4914 is obtained in > 80% yield from 2,4-lutidine and 2,4-dimethyl-6-(trimethylsilylmethyl)pyridine, b.p. 95 °C/12 mmHg, $n_D(20)$ 1.4868, in > 80% yield from *sym*-collidine.

c) Dimerizing coupling with iodine

To a vigorously stirred solution of 0.1 mol of 2-picolyllithium in ~ 70 ml of hexane and 80 ml of THF is added over 10 min a solution of 0.06 mol (15.2 g) of iodine in 60 ml of THF. During this addition the temperature is kept between -40 and -60 °C. A light-brown suspension is formed. After the addition, the cooling bath is removed and the temperature allowed to rise to -10 °C. A solution of 20 g of $Na_2S_2O_3$ in 150 ml of water is then added with vigorous stirring. After separation of the layers, the aqueous layer is extracted ten times with Et_2O. The combined organic solutions are dried over potassium carbonate and subsequently concentrated in vacuo; the last traces of volatile compounds are removed at ~ 0.5 mmHg (bath temperature not higher than 40 °C). The remaining brown solid is crystallized from a 5:1 mixture of pentane and Et_2O. The crystalline 2,2'-*bis*-picolyl, is obtained in ~ 70% yield.

d) β-hydroxyalkylation with oxirane

A mixture of 0.15 mol (6.6 g) of oxirane and 40 ml of Et_2O is added over 15 min to a solution of 0.1 mol of 2-picolyllithium in THF and hexane prepared as described above. During this addition, the temperature of the reaction mixture is kept between -20 and -5 °C. After an additional 15 min (at 0 °C), a solution of 5 g of water in 30 ml of THF is added with vigorous stirring over a few s. The light-yellow suspension (LiOH + H_2O) is very cautiously poured on a sintered-glass funnel (G-2) (connected to the water aspirator) which has been covered with a 1-cm thick layer of anhydrous sodium sulfate. After suction filtration, the solid is rinsed well with Et_2O. The clean filtrate is concentrated in vacuo and the remaining liquid distilled to give the alcohol py-$(CH_2)_3OH$, b.p. 115 °C/0.2 mmHg, $n_D(20)$ 1.5280 in 83% yield. The alcohol dissolves very well in water and over twenty extractions with Et_2O are necessary for a > 90% recovery from the aqueous solution.

e) Reaction with acetaldehyde

A mixture of 0.12 mol (5.3 g) of freshly distilled acetaldehyde and 20 ml of THF is added in one portion to a solution of 0.1 mol of 2-picolyllithium (prepared as described above) in THF and hexane cooled at -60 °C. After warming the reaction mixture to 0 °C, a mixture of 5 g of water and 30 ml of THF is added with vigorous stirring. The carbinol, b.p. 100 °C/0.2 mmHg, $n_D(20)$ 1.5152, is isolated in 76% yield in the manner described under d.

f) Reaction with ethyl cyanide
(compare Büchi J et al. (1962) Helv Chim Acta 45:729)

To a solution of 0.1 mol of 2-picolyllithium (prepared as described above) is added at 0 °C 0.12 mol (6.6 g) of ethyl cyanide. The temperature rises (without cooling bath) over a few min to ~ 20 °C. The orange-brown suspension is stirred for an additional 15 min at 20 °C, during which period lumps are formed. A mixture of 40 ml of 30% aqueous hydrochloric acid and 20 ml of water is then added at 0 °C and stirring is continued until the solid material has dissolved completely. The organic upper layer is almost colourless. The pH of the orange aqueous layer is brought to > 10 by addition of a 50% aqueous solution of potassium hydroxide, after which 15 extractions with small portions of chloroform are carried out. After drying the extracts over potassium carbonate, the solvent is removed in vacuo and the remaining yellow liquid distilled to give the ketone py-CH$_2$—COC$_2$H$_5$ (partly present in the enol form), b.p. 115 °C/12 mmHg, n$_D$(20) 1.5162, in 71% yield.

4.3 Regiospecific Generation of 4-Metallomethylpyridines in Organic Solvents

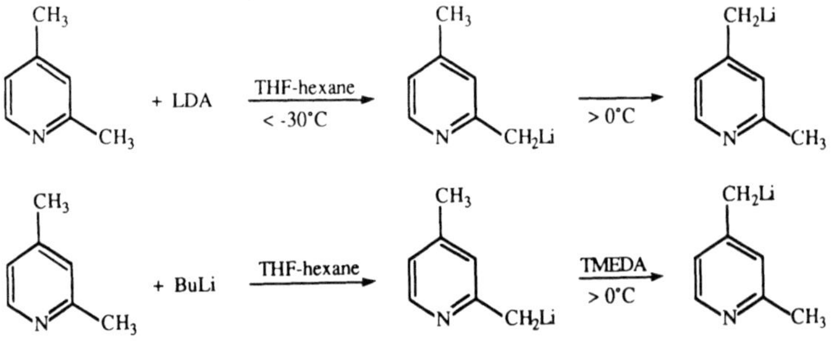

Apparatus: p. 24, Fig. 1; 500 ml

Scale: 0.1 molar

Introduction

As shown in Exp. 1, derivatives of pyridine having methyl groups on carbon atoms 2 and 4 are converted into 4-metallomethylpyridines upon treatment with alkali amides in liquid ammonia. Although these intermediates are formed almost instantaneously and with 100% selectivity, the transient occurrence of intermediates with the metal on the 2-CH$_3$ group cannot be excluded. If the metallation is carried out in Et$_2$O or THF with BuLi or LDA, at a low temperature, the 2-lithiomethyl-pyridines are generated with a very high regioselectivity and their stability is sufficient to permit a wide variety of regiospecific derivatization reactions (see Exp. 2). This kinetic stability can be ascribed to internal CH$_2$Li---N coordination (see Sect. 2, Ref. [2]). Transformation into the 4-lithiomethyl derivatives can be effected by compounds that can protonate this intermediate and whose kinetic acidity is comparable to that of the methyl groups in the pyridine derivative. The metallated compound formed by this proton exchange can subsequently abstract a proton from the 4-methyl group to give the thermodynamically more stable 4-lithiomethyl derivative. Secondary aliphatic amines such as diisopropylamine can function as proton-transferring agents. Thus, reaction of 2,4-dimethylpyridine with LDA at temperatures below $-30\,°C$ gives predominantly or exclusively the 2-lithiomethyl derivative as appears after quenching with an electrophile that reacts sufficiently fast at this low temperature. If the solution of 2-lithiomethylpyridine is warmed up to room temperature, a suspension is gradually formed. After about half an hour (at $20\,°C$) the isomerization to the 4-lithiomethylpyridine is complete.

CH₃ ring with CH₂Li (2-position) ⇌ (HDA / LDA) CH₃ ring with CH₃ (2-position) → (LDA) CH₂Li (4-position) ring with CH₃ (2-position)

Transformation of 2-lutidyllithium into the 4-isomer can also be achieved by adding a chelating reagent such as TMEDA or HMPT, which destabilizes 2-lutidyllithium by competitive coordination with the metal ion. Even simple warming may cause partial formation of 4-lutidyllithium. For relevant literature the reader should consult Ref. [2] of Sect. 2.

Procedure

1. Lithiation with LDA

2,4-Lutidine (0.11 mol, 11.8 g, dried over machine-powdered KOH and subsequently filtered) is added in one portion to a solution of 0.1 mol of LDA (Vol. 1, Chap. I) in ~ 70 ml of THF and ~ 66 ml of hexane, cooled at $-40\,°C$. The cooling bath is removed and the temperature allowed to rise. Above $0\,°C$ suspended material is formed in the red solution. After stirring for 30 min at $+20\,°C$, the formation of 4-lithio-lutidine is considered complete. 4-Methylpyridine can be metallated under similar conditions. With BuLi the metallation does not proceed cleanly (see under 2).

2. Lithiation with BuLi · TMEDA

THF (80 ml) is added to a solution of 0.1 mol of BuLi in 66 ml of hexane, cooled below $-30\,°C$. 2,4-Lutidine (0.11 mol, 11.8 g, dried over machine-powdered KOH) is added over 10 min with cooling between -30 and $-40\,°C$. TMEDA (0.1 mol, 11.6 g, dried over machine-powdered KOH) is then added in one portion and the red solution is stirred at $25\,°C$: after 5 to 10 min the solution becomes turbid and somewhat later two phases are formed (when stirring is stopped temporarily, two layers are visible). After 40 min (at $25\,°C$) the formation of 4-lutidyllithium is complete.

Treatment of 4-methylpyridine with BuLi gives rise to some addition across the C=N bond.

a) Trimethylsilylation
Chlorotrimethylsilane (0.11 mol, 11.9 g) is added in one portion at $-60\,°C$ to the solution of 4-lutidyllithium. The cooling bath is removed and stirring is continued for 15 min, then 100 ml of ice water is added and the product, 2-methyl-4-trimethylsilylmethylpyridine, b.p. $94\,°C/12$ mmHg, $n_D(22)$ 1.4949, is isolated in $> 80\%$ yield as described in Exp. 2.

b) Reaction with dimethyl disulfide
The solution of 4-lutidyllithium is poured (or transferred by a syringe with a needle of 1 mm diameter) over 5 min into a vigorously stirred mixture of 0.3 mol (28.2 g) of dimethyl disulfide and 80 ml of Et_2O. During this addition, the temperature is kept between -30 and $-70\,°C$. The product, 2-methyl-4-methylthiomethylpyridine, b.p. $104\,°C/12$ mmHg, $n_D(20)$ 1.5629 is isolated in $\sim 80\%$ yield in the manner described in Exp. 2.

4-Methylthiomethylpyridine, b.p. $110\,°C/12$ mmHg, $n_D(20)$ 1.5686, is obtained similarly (yield $\sim 80\%$) from 4-methylpyridine, LDA and CH_3SSCH_3.

c) Reaction with acetaldehyde
Addition at $-60\,°C$ of a 20% excess of freshly distilled acetaldehyde to the solution of 4-lutidyllithium, followed by a work-up as described in Exp. 2d, gives the carbinol, b.p. $\sim 100\,°C/0.2$ mmHg, in $\sim 75\%$ yield.

4.4 Metallation of 3-Methylpyridine

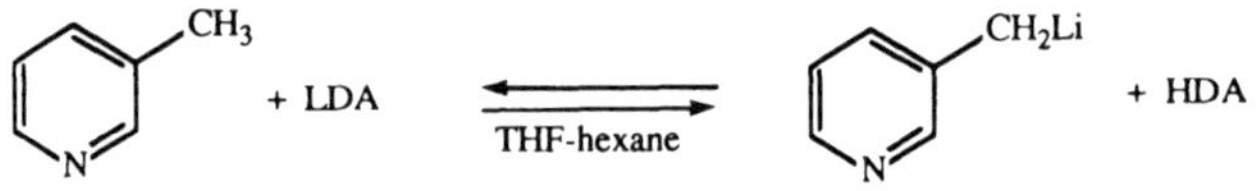

Subsequent alkylation with butyl bromide.

Apparatus: p. 24, Fig. 1; 11

Scale: 0.2 molar

Introduction

Since 3-methylpyridine is less acidic than diisopropylamine (see Sect. 1), the lateral metallation with LDA is expected to give an equilibrium mixture of 3-lithiomethylpyridine and LDA. The position of this equilibrium may depend to some extent upon the solvent: addition of the dipolar aprotic HMPT as a co-solvent (compare Sect. 2, Ref. [3]) presumably will cause a shift in favour of the metallated pyridine, since especially the N—Li bond in LDA is destabilized. In spite of the incomplete metallation, good yields of the homologues may be obtained when adding an alkyl bromide. This is more reactive towards the lithiated pyridine than towards LDA, so that the equilibrium can shift to the side of 3-lithiomethylpyridine. A similar situation may arise during reactions with oxirane and homologues. Other electrophiles, such as chlorotrimethylsilane, disulfides (RSSR), and most carbonyl compounds react at comparable rates with LDA and the lithiated pyridine (or faster even with LDA) and the results of these reactions are poor. Our attempts to achieve a complete metallation by using BuLi·t-BuOK in THF or in combination with TMEDA were not successful, the main reaction probably being addition of the base across the C=N bond.

Procedure (compare Sect. 2, Ref. [3])

3-Methylpyridine (0.22 mol, 20.5 g, dried over machine-powdered KOH and subsequently filtered) is added over a few s to a solution of 0.2 mol of LDA (see Vol. 1, Chap. I) in 180 ml of THF and 132 ml of hexane, cooled at − 20 °C. The cooling bath is removed and the temperature allowed to rise to + 10 °C. After an additional half hour the orange suspension is cooled to − 40 °C and 0.3 mol (41.1 g) of butyl bromide is added in one portion. The cooling bath is then removed and the temperature allowed to rise. Above − 15 °C the heating effect is clearly observable. Some cooling may be necessary to keep the temperature between 20 and 30 °C. After stirring for an additional 15 min (at 30–35 °C), 100 ml of water is added with vigorous stirring. The aqueous layer is extracted twice with Et$_2$O. The combined organic solutions are dried over potassium carbonate, then concentrated *in vacuo* and the remaining liquid is carefully distilled. 3-Pentylpyridine, b.p. 100 °C/12 mmHg, n$_D$(20) 1.4911, is obtained in an excellent yield.

Reaction of the solution of 3-lithiomethylpyridine with acetaldehyde, dimethyl disulfide (inversed-order) and chlorotrimethylsilane gave the products in poor (< 10%) yields.

4.5 Lithiation of 2-Methylthiazoline

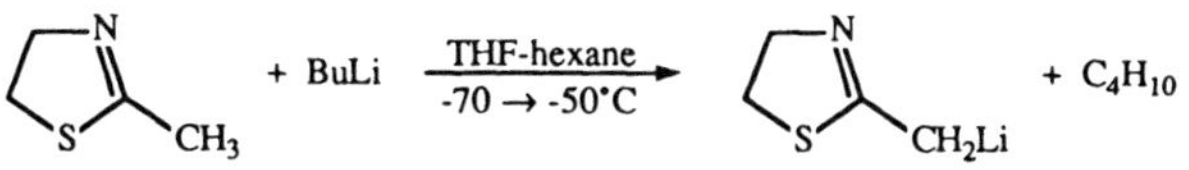

Functionalization with *n*-butyl bromide, chlorotrimethylsilane and dimethyl disulfide.

Apparatus: p. 24, Fig. 1; 500 ml

Scale: 0.1 molar

Introduction: See Sects. 1 and 3

Procedure

2-Methylthiazoline (0.1 mol, 10.1 g, commercially available) is added in one portion to a solution of 0.105 mol of BuLi in 70 ml of hexane and 80 ml of THF, cooled at − 70 °C. The temperature rises within a few s to ~ − 50 °C. After an additional 5 min (at − 50 °C), the thin and almost white suspension is cooled to − 80 °C and butyl bromide (0.15 mol, 20.5 g) or freshly distilled chlorotrimethylsilane (0.11 mol, 11.9 g) is added in one portion with vigorous stirring. The cooling bath is removed immediately. The reaction with Me₃SiCl is very fast and a light-yellow solution is formed. After an additional 10 min (at − 40 °C), the work-up (see below) is carried out. The butylation proceeds much more slowly and the temperature is allowed to rise to + 10 °C prior to the work-up (see below). The methylthiolation is carried out by transferring (over 5 min) the solution of 2-lithiomethylthiazoline (cooled at − 30 °C) by syringe into a vigorously stirred mixture of 0.3, mol (28.2 g) of dimethyl disulfide and 60 ml of THF while cooling below − 30 °C. No additional time is necessary (note 1). In all cases most of the solvent is first removed under reduced pressure (rotary evaporator), then 100 ml of water is added and the product is isolated by extraction (twice) with Et₂O, drying over potassium carbonate, removal of the solvent *in vacuo*, and distillation through a 30-cm Vigreux column. 2-Pentylthiazoline, b.p. 98 °C/15 mmHg, $n_D(20)$ 1.4956; 2-trimethylsilyltmethyl-thiazoline, b.p. 87 °C/15 mmHg, $n_D(20)$ 1.4974, and 2-methylthiomethylthiazoline, b.p. 120 °C/15 mmHg, $n_D(20)$ 1.5708, are obtained in greater than 75% yields.

Notes

1. If dimethyl disulfide is added dropwise, the desired product is obtained in low yield. The main product is the *bis*-methylthio derivative, formed by proton transfer from the initial product and subsequent introduction of a second methylthio group.

4.6 Lithiation of 2,4,4,6-Tetramethyl-5,6-dihydro-1,3-oxazine

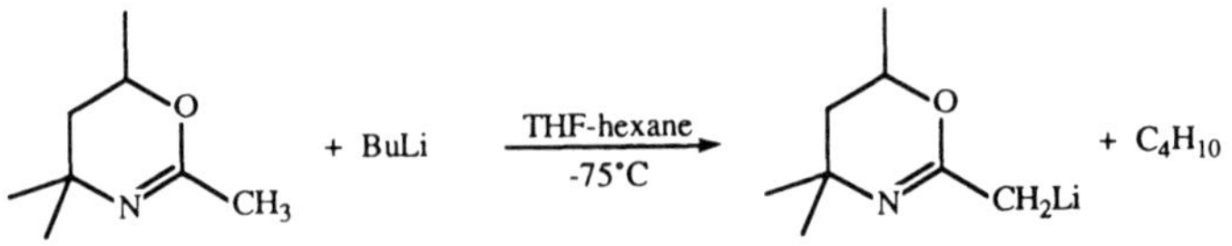

Derivatization reactions with butyl bromide and trimethylchlorosilane.

Apparatus: p. 24, Fig. 1; 500 ml

Scale: 0.1 molar

Introduction

In their article, Meyers and co-workers (see Sect. 1, Ref. [4]) give a general procedure for the lithiation of the 2-methyldihydrooxazine derivative. A 10% molar excess of BuLi is added over 1 hour at $-78\,^\circ$C to a mixture of the substrate and THF, after which the reaction mixture is stirred for an additional 1 h at this temperature. The obtained suspension of the 2-lithiodihydrooxazine is then allowed to react with a large variety of electrophiles. In our procedure the order of addition is reversed and the *substrate* is used in excess. The presence of BuLi during functionalization reactions could lead to introduction of a second substituent, especially in reactions with chlorotrimethylsilane and disulfides. In order to obtain optimal yields, the temperature during the reactions with electrophiles has to be maintained below -50 or $-60\,^\circ$C. At higher temperatures an irreversible ring opening with formation of $H_2C\!=\!C\!=\!N\!-\!C\!-\!CH_2\!-\!C\!-\!OLi$ sets in.

Meyers et al. dry the starting compound over anhydrous potassium carbonate. We found subsequent treatment of the ethereal solution of the dihydrooxazine with machine-powdered KOH to be necessary.

Procedure

A mixture of 0.11 mol (15.5 g) of the dihydrooxazine (Ref. [4] of Sect. 1) and 20 ml of THF is added over a few min to a solution of 0.1 mol of BuLi in 66 ml of hexane and 70 ml of THF. During this addition the temperature of the solution is kept between -70 and $-80\,^\circ$C by occasional cooling in a bath with liquid nitrogen. The mixture is stirred for 1 h at $\sim -75\,^\circ$C, during which period a fine, light-yellow suspension may be formed. Butyl bromide (0.15 mol, 20.5 g) or freshly distilled chlorotrimethylsilane (0.13 mol, 14.1 g) is then added over 10 min. After these additions, the temperature is allowed to rise gradually over 15 min to $-60\,^\circ$C. In the case of the silylation, the cooling bath is then removed and after an additional 15 min, the greater part of the solvent is removed under reduced pressure using a rotary evaporator (Note 1). The reaction with butyl bromide is completed by gradually raising the temperature over 45 min to $0\,^\circ$C, then the solvent is removed as described for the silylation.

In both cases, the work-up is carried out by successively adding 100 ml of Et_2O and 100 ml of ice water to the remaining sirupy liquid. After vigorous shaking and separation of the layers, the aqueous phase is extracted three times with Et_2O. The unwashed ethereal solutions are dried over anhydrous potassium carbonate and subsequently concentrated in vacuo. Careful fractionation of the remaining liquid through a 40-cm Vigreux column gives the butyl derivative, b.p. $86\,^\circ$C/12 mmHg, $n_D(20)$ 1.4182 (yield 77%) and the silylation product, b.p. $96\,^\circ$C/12 mmHg, $n_D(20)$ 1.4430 (yield 85%).

Notes

1. The excess of Me$_3$SiCl is also removed during evacuation: the presence of Me$_3$SiCl during the aqueous work-up may give rise to cleavage of the C—Si bond with back-formation of the 2-methyldihydrooxazine (see Chap. I, Sect. 10).

4.7 Dimetallation of 2,6-Dimethylpyridine

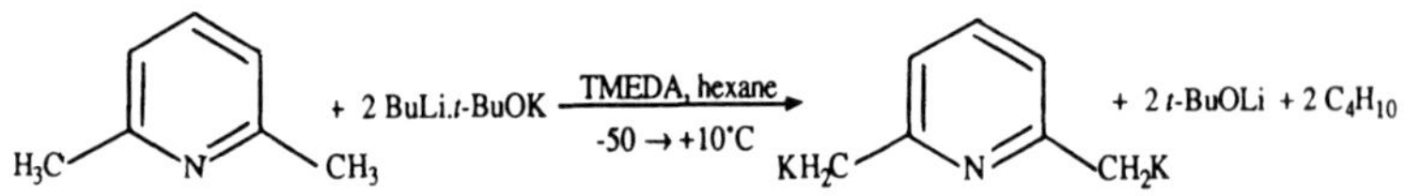

Subsequent reaction with chlorotrimethylsilane.

Apparatus: p. 24, Fig. 1; 500 ml

Scale: 0.05 molar

Introduction

t-BuOK (0.13 mol, 14.6 g, powdered in a dry atmosphere), hexane (50 ml), and TMEDA (0.13 mol, 15.1 g) are placed in the flask. A solution of 0.13 mol of BuLi in 84 ml of hexane is added over 5 min with cooling below − 40 °C and stirring at a moderate rate. After an additional 15 min (at − 35 °C), the mixture is cooled to − 50 °C and 0.05 mol of 2,6-lutidine (5.4 g) dissolved in 20 ml of hexane is added over a few s. The temperature is then allowed to rise gradually over 1 h to + 10 °C. THF (80 ml) is added and after an additional 15 min (Note 1), the orange suspension is cooled to − 70 °C. Freshly distilled chlorotrimethylsilane (0.13 mol, 14.1 g) is added in one portion with vigorous stirring, after which the cooling bath is removed and the temperature allowed to rise to 15 °C. The greater part of the solvent is then removed under reduced pressure using a rotary evaporator. Et$_2$O (100 ml) and 100 ml of ice water are added to the residue. After shaking and separation of the layers, three extractions with Et$_2$O are carried out. The unwashed organic solutions are dried over potassium carbonate and then concentrated in vacuo. Careful fractional distillation gives, after a small forerun of the monosilylated derivative, the *bis*-trimethylsilyl derivative, b.p. 122 °C/12 mmHg, in ~80% yield. The compound solidifies upon standing at room temperature.

Notes

1. The excess of BuLi·*t*-BuOK reacts with THF to give ethene and H$_2$C=CHOK.

5 Organic Syntheses Procedures

Kofron WG, Baclawski LM (1972) Org Synth 52:75

Walter LA (1955) Org Synth, Coll 3:757:

Woodward RB, Kornfeld EC (1955) Org Synth Coll 3:413

Politzer IR, Meyers AI (1971) Org Synth 51:24

$$Ox—CH_2Ph \xrightarrow[\text{2. Br(CH}_2)_4\text{Br}]{\text{1. BuLi}} Ox—CH(Ph)(CH_2)_4Br \xrightarrow{\text{BuLi}}$$

Chapter VII
Metallation of Aldimines and Ketimines

1 Introduction

Directed functionalizations at the α-carbon atom of aldehydes and ketones have been carried out via reactions of their nitrogen derivatives, including aldimines, ketimines, hydrazines, and oximines. The general procedure involves metallation of the nitrogen derivative, subsequent reaction with an electrophile and final conversion of the obtained derivative into the derivative of the starting aldehyde or ketone. The reaction sequence is depicted in the following general scheme:

$$-\underset{\underset{H}{|}}{C}-C{=}O \xrightarrow{H_2NY} -\underset{\underset{H}{|}}{C}-C{=}N-Y \xrightarrow{base} -\underset{\underset{M}{|}}{C}-C{=}N-Y$$

$$\xrightarrow{E^+} -\underset{\underset{E}{|}}{C}-C{=}N-Y \xrightarrow[\text{other reagent}]{H^+,H_2O \text{ or}} -\underset{\underset{E}{|}}{C}-C{=}O$$

$$(Y = t\text{-}C_4H_9, c\text{-}C_6H_{11}, N(\text{alkyl})_2, OLi, OR)$$

The imines (Schiff's bases) have found most general application. Their preparation as well as the conversion of their derivatization products into aldehydes or ketones can be carried out under mild conditions and with satisfactory yields using extremely simple auxilliary reagents.

The use of the metallated nitrogen derivatives has some distinct advantages over that of their synthetic equivalents, the enolates, which are connected with their higher basicity and, therefore, higher reactivity towards electrophiles. Enolates derived from aldehydes and ketones are not very reactive towards alkyl halides. As a consequence, proton transfer from the initial alkylation products to the enolate and subsequent reaction of the newly generated enolate with alkyl halide resulting in di- and polyalkylation products, seriously compete with the intended reactions. Similar complications arise during attempts to introduce an SR group in the α-position of a carbonyl function using disulfides, sulfenyl halides, or other sulfenylating agents.

Metallated aldimines are useful substitutes for aldehyde enolates, which in general, are not easy to generate in a straight-forward manner. The facile transformation of aldehydes or ketones into the corresponding imines or hydrazines offers an excellent opportunity to introduce chiral nitrogen-containing groups. Using optically active imines and hydrazines, Meyers and Enders and their co-workers succeeded in synthesizing ketones with high optical purity [1–3].

The chemistry of the metallated nitrogen derivatives of aldehydes and ketones has been reviewed by Mukaiyama [4] and Whitesell–Whitesell [5].

In this chapter, some procedures with simple aldimines and ketimines are described. They show representative conditions for the metallation of these nitrogen derivatives, their subsequent reactions with a number of electrophiles and the final hydrolytic cleavage of the C=N bond with formation of derivatives of the starting aldehydes or ketones.

Literature

1. Meyers AI, Williams DR, White S, Erickson GW (1981) J Am Chem Soc 103:3088
2. Enders D, Eichenauer H, Baus U, Schubert H, Kremer KAM (1984) Tetrahedron 40:1345
3. Enders D, Kipphardt H, Fey P (1987) Org Synth 65:183
4. Mukaiyama T (1982) Org Reactions 28:203, John Wiley
5. Whitesell JK, Whitesell MA (1983) Synthesis 517

2 Conditions for the Metallation

The most usual reagent for the metallation of aldimines and ketimines and the corresponding hydrazines is LDA. The lithiations are generally carried out in mixtures of THF and hexane. The formation of the lithium derivatives, which proceeds rather smoothly at temperatures in the region -30 to $+10\,°C$, is visible by the appearance of a yellow colour. The use of BuLi cannot be recommended, since addition of this reagent across the C=N bond may seriously compete with the deprotonation.

As may be concluded from the high yields of some alkylation reactions performed in liquid ammonia, the interaction between aldimines and sodamide or potassium amide gives rise to predominant or complete metallation. With lithium amide the ionization equilibrium is probably strongly on the side of the reactants.

3 Experiments

All temperatures are internal, unless indicated otherwise.

All reactions are carried out in an atmosphere of inert gas, except those in liquid ammonia.

For general instructions concerning handling organolithium reagents, drying solvents, etc. see Vol. 1.

3.1 Lithiation of Aldimines and Ketimines with LDA

$$RCH_2CH{=}NR' + LDA \xrightarrow[-50\to+15°C]{THF\text{-}hexane} RCH{=}CHN(Li)R' + HDA$$

$$R = H, CH_3; R' = t\text{-Bu}, c\text{-}C_6H_{11}$$

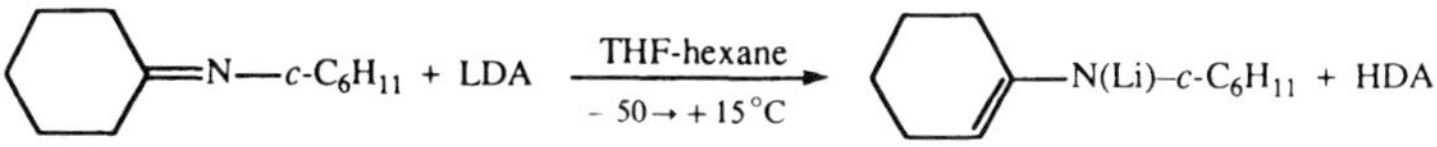

$$\xrightarrow[-50\to+15°C]{THF\text{-}hexane}$$

(Analogous for the cyclopentanone imine)

Apparatus: p. 24, Fig. 1; 11

Scale: 0.1 molar

Procedure

The imine (0.11 mol, Exp. 7 and Note 1) dissolved in 30 ml of THF is added in one portion to a solution of 0.1 mol of LDA (Vol. 1, Chap. I) in 60 ml of THF and 66 ml of hexane (Note 2), cooled at $-55°C$. The cooling bath is removed and the temperature allowed to rise to $+15°C$. After an additional 10 min (aldimines) or 30 min (ketimines), the light-yellow (aldimines) to dark-yellow (ketimines) solution is used for reactions with electrophiles (Note 3).

Notes

1. Especially $CH_3CH{=}N{-}t\text{-Bu}$ and $C_2H_5CH{=}N{-}t\text{-Bu}$ undergo decomposition during storage, even at $-20°C$, so that redistillation before use may be necessary.
2. If, instead of the imine $CH_3CH{=}N{-}t\text{-Bu}$, LDA is used in excess, a corresponding amount of dialkylation product is formed in the reactions with alkyl halides; the metallations of the other imines are carried out with a 5% molar excess of LDA.
3. The solutions of the lithiated imines are reasonably stable at $20–30°C$.

3.2 Alkylation of Lithiated Imines

$$RCH{=}CH{-}N(Li){-}t\text{-Bu} + R'X \xrightarrow{THF\text{-}hexane}$$

1

$$RR'CHCH{=}N{-}t\text{-Bu} + LiX$$

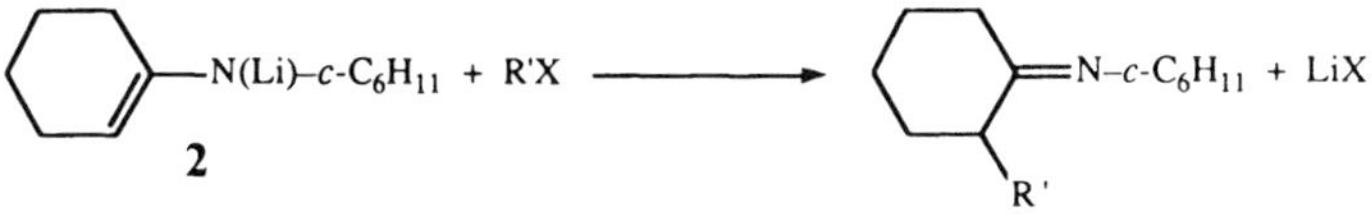

Apparatus: p. 24, Fig. 1; 11

Scale: 0.1 molar

Procedure

1. Reaction of 1, (R = H), with hexyl bromide and benzyl chloride

Hexyl bromide (0.13 mol, 21.4 g) or benzyl chloride (0.13 mol, 16.4 g) is added in one portion at $-20\,°C$ to the solution of **1**, R = H, after which the cooling bath is removed. In the reaction with hexyl bromide the heating effect becomes observable in the region of 0 to 5 °C, the benzylation becomes 'visible' above $-15\,°C$. In both cases the tempeature of the reaction mixture rises to $\sim 30\,°C$ over 10 to 15 min. The solutions are heated for an additional 40 min at 40 °C, after which they are cooled to 0 °C. Ice water (150 ml) is added with vigorous stirring. The unwashed (Note 1) organic layer and two ethereal extracts are dried over potassium carbonate and subsequently concentrated under reduced pressure. Careful distillation through a 30-cm Vigreux column gives the products: R = H, $R' = C_6H_{13}$, b.p. 90 °C/12 mmHg, $n_D(22)$ 1.4308 (yield ~ 85–90%) and R = H, $R' = CH_2Ph$, b.p. 112 °C/12 mmHg, $n_D(22)$ 1.4968 (yield ~ 80%).

2. Reaction of 1, R = CH₃, with bromoacetaldehyde diethylacetal
   (compare Normant H et al. (1976) Tetrahedron Lett 1379)

The solution of **1**, $R = CH_3$, is cooled to $-50\,°C$ and 30 ml of purified HMPT (see Vol. 1, Chap. I and p. 225) is added. The bromoacetal (0.15 mol, 30.0) is added in one portion at $-50\,°C$, then the cooling bath is removed. The temperature of the light-brown solution is allowed to rise to $+10\,°C$. After an additional 30 min of heating at 35–40 °C, the solution is cooled to 0 °C and 150 ml of ice water is added with vigorous stirring. The organic layer and four ethereal extracts are washed three times with a saturated aqueous solution of ammonium chloride, then dried over potassium carbonate and the remaining liquid is carefully distilled through a 30-cm Vigreux column. The product, b.p. 101 °C/12 mmHg, $n_D(22)$ 1.4312, is obtained in ~ 50% yield. Starting from **1**, R = H, the coupling product is obtained in only ~ 25% yield.

3. Reaction of 1, R = H with bromochloropropane

Bromochloropropane (23.6 g, 0.15 mol) is added in one portion at $-10\,°C$ to the solution of **1**, whereupon the cooling bath is removed. The temperature rises over a few min to over 30 °C but is kept between 30 and 35 °C by occasional cooling (Note 2). After an additional 30 min (at 30–35 °C), the solution is cooled to 0 °C and 150 ml of ice water is added with vigorous stirring. The organic layer and three ethereal extracts are dried over potassium carbonate and subsequently concentrated under reduced pressure, keeping the bath temperature below 40 °C. This work-up should be carried out without delay (note 2). The remaining liquid, a mixture of $Cl(CH_2)_4CH=N-t\text{-Bu}$, $Br(CH_2)_3Cl$ and some THF, is used for the conversion into the aldehyde (see Exp. 5).

4. Reaction of 2 with butyl bromide and allyl bromide

Butyl bromide (0.15 mol, 20.5 g) or allyl bromide (0.12 mol, 14.5 g) is added over a few seconds at -30 and $-80\,°C$, respectively, whereupon the cooling bath is removed. In the reaction with C_4H_9Br the temperature rises over 10 min to $25\,°C$, the reaction with $H_2C=CHCH_2Br$ is extremely fast as indicated by almost immediate disappearance of the yellow colour and a temperature rise to $\sim -15\,°C$ over a few s. The butylation is terminated by heating the solution for 30 min at $35\,°C$: a faint yellow colour remains. The products are isolated as described under 1. The yields of the undistilled products are greater than 80%. For their conversion into the ketones see Exp. 5.

Notes

1. Washing until the aqueous layer has become neutral involves the risk of hydrolysis of part of the products to the aldehydes.
2. The product has a low thermal stability due to the formation of the cyclic imonium salt. After standing for two days at room temperature the product has become very viscous.

3.3 Metallation of Aldimines by Sodamide in Liquid Ammonia

$$CH_3CH=N-t\text{-Bu} + NaNH_2 \xrightarrow[-33\,°C]{\text{liq. } NH_3}$$

$$NaCH_2CH=N-t\text{-Bu} + NH_3$$

Subsequent alkylation with hexyl bromide and cyclopentyl bromide.

Apparatus: 1-l, three-necked, round-bottomed flask, equipped with a dropping funnel, a mechanical stirrer and a gas-outlet (rubber stopper with a hole of ~ 7 mm)

Scale: 0.2 molar

Introduction

Although we had good results in some subsequent reactions with alkyl halides, the metallation of imines by alkali amides has a limited applicability. We assume, on the basis of our results, that the pK values of the imines are close to that of ammonia. Thus, the reaction with alkali amides in this solvent will give rise to certain equilibrium concentrations of the metallated imines, which depend upon the counterion (Li^+, Na^+, or K^+) and the imine (aldimine or ketimine, substituents). The low yields in the reactions of $CH_3CH=N-t$-Bu with lithium amide and of the cyclohexanone imine with sodamide indicate that these metallation equilibria are strongly on the side of the reactants. As may be concluded from the high yield in the alkylation with hexyl bromide, interaction between the simple imine $CH_3CH=N-t$-Bu and sodamide gives rise to a sufficiently high equilibrium concentration of the imine anion (compare Chap. VI, Exp. 1). With $C_2H_5CH=N-t$-Bu the ionization with sodamide is less complete.

Procedure

The freshly distilled imine (0.22 mol, 21.8 g, Exp. 7, see also Exp. 1, Note 1) is added over a few min to a suspension of 0.2 mol of sodamide in 350 ml of liquid ammonia (see Vol. 1, Chap. I). A greenish solution is formed. After an additional 2 min, 0.2 mol (33.0 g) of hexyl bromide or 0.2 mol (30.0 g) of cyclopentyl bromide is added dropwise over 10 min with vigorous stirring. The reactions are very vigorous and are finished shortly after the addition of the halides. The equipment is then removed and the greater part of the ammonia is removed by placing the flask in a water bath at $\sim 35\,°C$, stirring being continued. Et_2O (150 ml) is added to the remaining grey slurry, after which the flask is placed in an ice-water bath. Ice water (150 ml) is added over 1 min with vigorous stirring. The aqueous layer is extracted three times with Et_2O. The unwashed organic solutions are dried over potassium carbonate and subsequently concentrated in vacuo. Careful distillation of the remaining liquid gives $C_6H_{13}CH_2CH=N-t$-Bu, b.p. 90 °C/12 mmHg, $n_D(20)$ 1.4316 (yield 82%) and cyclopentyl-$CH_2CH=N-t$-Bu, b.p. 78 °C/12 mmHg, $n_D(20)$ 1.4500 (yield 45%).

3.4 Trimethylsilylation and Methylthiolation of Lithiated Imines

$$RCH=CH-N(Li)-t\text{-Bu} + Me_3SiCl \xrightarrow{\text{THF-hexane}} RCH(SiMe_3)CH=N-t\text{-Bu}$$

$$+ CH_3SSCH_3 \longrightarrow RCH(SCH_3)CH=N-t\text{-Bu}$$

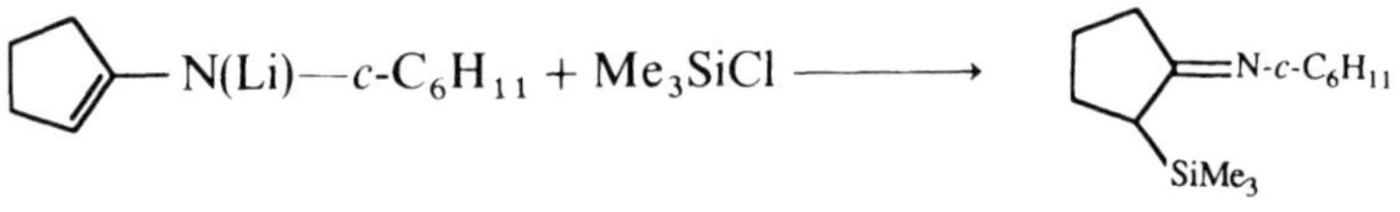

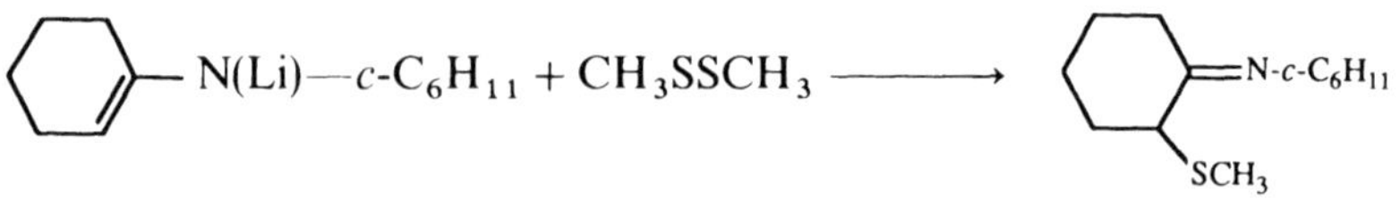

Apparatus: p. 24, Fig. 1; 11

Scale: 0.1 molar

1. Trimethylsilylation

Freshly distilled chlorotrimethylsilane (0.11 mol, 11.9 g) is added in one portion to the solution of the lithiated imine (Exp. 1) cooled at $-50\,°C$. The cooling bath is removed and the temperature allowed to rise to $+5\,°C$. The greater part of the solvent is then removed under reduced pressure, using a rotary evaporator (bath temperature $<40\,°C$). Cold Et_2O (100 ml, $0\,°C$) and ice water (100 ml) are succesively added to the residue. After vigorous shaking, the layers are separated and the aqueous layer is extracted twice with Et_2O. The organic solutions are dried over potassium carbonate and subsequently concentrated in vacuo. Careful distillation of the remaining liquid affords the following compounds in $>80\%$ yields:

$Me_3SiCH_2CH{=}N{-}t$-Bu, b.p. $46\,°C/12$ mmHg, $n_D(20)$ 1.4263;
$CH_3CH(SiMe_3)CH{=}N{-}t$-Bu, b.p. $60\,°C/12$ mmHg, $n_D(20)$ 1.4330;

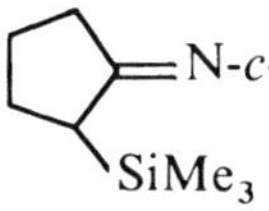

$=N$-c-C_6H_{11}, b.p. $\sim 100\,°C/1$ mmHg, $n_D(20)$ 1.4845
$SiMe_3$

2. Methylthiolation

a) The solution of $H_2C{=}CH(Li)N{-}t$-Bu is added (by syringe or from a dropping funnel) over 5 min to a vigorously stirred mixture of 0.2 mol (18.8 g, excess) of dimethyl disulfide and 60 ml of Et_2O, while keeping the temperature below $-40\,°C$. Water (150 ml) is then added with vigorous stirring. The aqueous layer is extracted twice with Et_2O. The (unwashed) organic solutions are dried over potassium carbonate and subsequently concentrated in vacuo. Careful distillation of the remaining liquid through a 30-cm Vigreux column gives $CH_3SCH_2CH{=}N{-}t$-Bu, b.p. $70\,°C/12$ mmHg, $n_D(20)$ 1.4733, in $\sim 75\%$ yield. A considerably lower yield is obtained if CH_3SSCH_3 is added dropwise to the solution of the lithiated aldimine. The residue of the distillation is considerable and consist mainly of $(CH_3S)_2CHCH{=}N{-}t$-Bu.

b) Dimethyl disulfide (0.15 mol, 14.1 g) is added in one portion to a vigorously stirred solution of the cyclohexanone imine cooled at $-70\,°C$. Five minutes after this addition, 100 ml of ice water is added and the product is isolated as described above. The undistilled product is used for the conversion into 2-methylthiocyclohexanone (Exp. 5).

3.5 Conversion of Imines into Aldehydes and Ketones

$$RCH_2CH{=}N{-}t\text{-}Bu + H_2O + HCl \xrightarrow[20\to 80\,°C]{pH\ 5\text{-}6}$$

$$RCH_2CH{=}O + t\text{-}BuNH_2 \cdot HCl$$

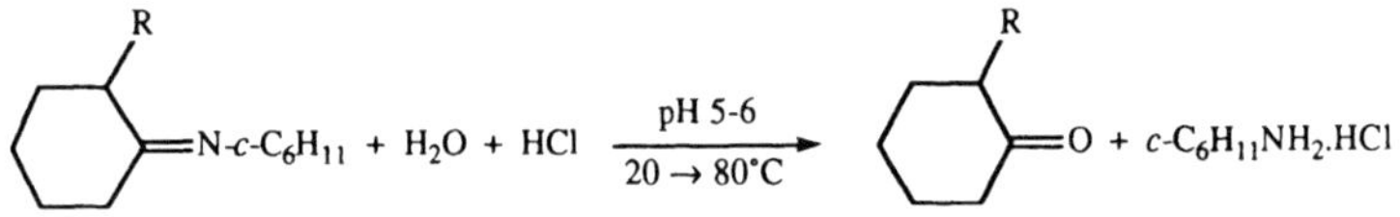

(Analogous for five-membered ring.)

Apparatus: 1 l, three-necked, round-bottomed flask, equipped with a dropping funnel, a mechanical stirrer and a thermometer.

Scale: ~0.2 molar

Introduction

Aldimines and ketimines react with an excess of aqueous acid to give solutions of iminium salts $\overset{+}{>}C{=}NHR$. When the pH is sufficiently low (<4), these solutions remain unchanged at room temperature. Upon heating, aldehydes or ketones are liberated. This hydrolytic process is accelerated by increasing the pH, the optimal rate being reached at pH ~6 [1].

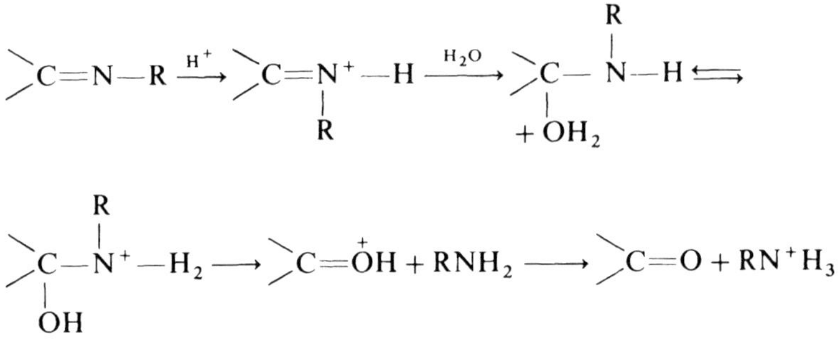

In the usual procedure, the hydrolytic cleavage is carried out with oxalic acid, sometimes very diluted [2]. We found it more convenient to dissolve the imine in a cold (~0 °C) aqueous solution of a 10 to 20 mol% excess of hydrochloric acid. Impurities (if present) may be removed by extracting the cold acidic solution (pH < 2) with Et$_2$O or pentane. The pH of the aqueous solution is then brought at 5.5 to 6 by controlled addition of an aqueous solution of potassium carbonate, after which the solution is heated.

Procedure

The distilled or crude functionalization product (~ 0.2 mol, Exp. 2-1, 2-3, 2-4, 4-2b) is added over a few min to a vigorously stirred solution of 0.25 mol of hydrochloric acid in 150 ml of water, kept at 0 °C by cooling in a dry ice-acetone mixture. The dropping funnel is rinsed with a small amount of Et_2O. The cold aqueous solution is then immediately extracted twice with pentane, the pentane layers being discarded. A concentrated aqueous solution of potassium carbonate is then added dropwise with stirring until the pH of the aqueous layer has become 5.5 to 6. The mixture is heated for 30 min at 70 °C, then cooled to room temperature and extracted with Et_2O. After drying the extract over $MgSO_4$, the solvent is removed in vacuo and the remaining liquid distilled. The following aldehydes and ketones have been prepared:

$C_6H_{13}CH_2CH{=}O$, b.p. 58 °C/12 mmHg, $n_D(24)$ 1.4185, 88% yield;
$Cl(CH_2)_4CH{=}O$, b.p. 68 °C/12 mmHg, $n_D(21)$ 1.4457, 70% overall yield;
2-allylcyclohexanone, b.p. 82 °C/12 mmHg, 86% overall yield;
2-butylcyclohexanone, b.p. 100 °C/15 mmHg, $n_D(22)$ 1.4538, 85% overall yield;
2-methylthiocyclohexanone, b.p. 96 °C/15 mmHg, $n_D(20)$ 1.5103; 80% overall yield.

Literature

1. Hine J, Craig Jr JC, Underwood JG, Via FA (1970) J Am Chem Soc 92:5194
2. Dauben WG, Beasley GH, Broadhurst MD, Muller B, Peppard DJ, Pesnelle P, Suter C (1975) J Am Chem Soc 97:4973

3.6 Synthesis of α,β-Unsaturated Aldehydes from Trimethylsilylated Aldimines

$$CH_3CH(SiMe_3)CH{=}N{-}t\text{-}Bu + LDA \xrightarrow[-40 \to +20\,°C]{\text{THF-hexane}}$$

$$CH_3C(SiMe_3){=}CH{-}N(Li){-}t\text{-}Bu + HDA \xrightarrow[\text{(Peterson elim.)}]{R'R''C{=}O}$$

$$R'R''C{=}C(CH_3)CH{=}N{-}t\text{-}Bu \xrightarrow[\text{pH 5.5}]{H^+,\,H_2O} R'R''C{=}C(CH_3)CH{=}O$$

$$(R'R''C{=}O = \text{cyclohexanone or } PhCH{=}O)$$

Apparatus: p. 24, Fig. 1; 11

Scale: 0.1 molar

Introduction

Coupling of lithiated aldimines with ketones or aldehydes gives the expected carbinols in good yields. Upon heating with water at pH $\sim$ 6, the aldols are formed (see Exp. 5 for the procedure). Subsequent treatment with acid should give the α,β-unsaturated aldehyde. However, in many cases this dehydration requires rather drastic conditions and does not proceed cleanly: appreciable amounts of the starting aldehyde or ketone are sometimes isolated [1]. Remarkably, Büchi and Wüest [2] reported the formation of nuciferal $RCH=C(Ch_3)CH=O$ (R = p-CH_3— C_6H_4—$CH(CH_3)CH_2CH_2$—) in good yield by treatment of the aldol $RCH(OH)CH(CH_3)CH=O$ (not isolated) in ethereal solution with cold dilute oxalic acid. A more satisfactory formation of α,β-unsaturated aldehydes is achieved by coupling carbonyl compounds with lithiated imines containing an α-phosphorus substituent [3–6] or α-trimethylsilyl group [7] and subsequent treating the product with water.

Procedure

Cyclohexanone (0.1 mol, 9.8 g) or benzaldehyde (0.1 mol, 10.6 g) is added in one portion to a solution of 0.12 mol (excess) of the lithiated trimethylsilylaldimine [prepared by adding the imine (0.12 mol) to a solution of 0.12 mol of LDA in 80 ml of THF and 80 ml of hexane cooled at $-40\,^\circ$C and subsequently keeping the yellowish-brown solution for 2 min at $+20\,^\circ$C, compare Exp. 1] cooled to $-75\,^\circ$C. The temperature rises within a few s to $-20\,^\circ$C or higher. The Peterson elimination is completed by stirring the reaction mixture for an additional 15 min at $+25\,^\circ$C. The reaction mixture is then cooled again to $-30\,^\circ$C and a solution of 0.4 mol of hydrochloric acid in 200 ml of ice water is added with vigorous stirring, keeping the temperature between -5 and $+5\,^\circ$C. After separation of the layers, the aqueous layer is extracted twice with a 1:1 mixture of Et_2O and pentane, the extracts being discarded. The aqueous layer is then brought at pH 5.5 by controlled addition of a concentrated solution of potassium carbonate and subsequently heated for 30 min at $75\,^\circ$C. After cooling to room temperature, four extractions with Et_2O are carried out. The combined extracts are washed once with 1 N hydrochloric acid and subsequently dried over $MgSO_4$. Careful distillation through a 20-cm Vigreux column gives the aldehydes $PhCH=C(CH_3)CH=O$, b.p. $\sim 70\,^\circ$C/0.1 mmHg, and $(CH_2)_5C=C(CH_3)CH=O$, b.p. $107\,^\circ$C/12 mmHg, $n_D(20)$ 1.5078, in $\sim 70\%$ yield.

Literature

1. Unpublished observations in the author's laboratory
2. Büchi G, Wüest H (1969) J Org Chem 34:1122
3. Nagata W, Hayase Y (1968) Tetrahedron Lett 4359
4. Nagata W, Hayase Y (1969) J Chem Soc (C) 460
5. Portnoy NA, Morrow CJ, Chattha MS, Williams Jr JC, Aguilar AM (1971) Tetrahedron Lett 1401; Chattha MS, Aguilar AM (1971) Tetrahedron Lett 1419
6. Chattha MS, Aguilar AM (1971) J Org Chem 36:2892
7. Corey EJ, Enders D, Bock MG (1976) Tetrahedron Lett 7

3.7 Preparation of Aldimines and Ketimines

$$RCH{=}O + R'NH_2 \xrightarrow[\text{K}_2\text{CO}_3]{0 \to 30\,^{\circ}\text{C}} RCH{=}NR' + H_2O \tag{1}$$

$$R{=}CH_3 \text{ or } C_2H_5; \ R' = t\text{-Bu or } c\text{-}C_6H_{11}$$

$$(CH_2)_nC{=}O + c\text{-}C_6H_{11}NH_2 \longrightarrow (CH_2)_nC{=}N{-}c\text{-}C_6H_{11} + H_2O \tag{2}$$

$$(n = 4 \text{ or } 5)$$

Apparatus: for reaction (1): 1-l, round-bottomed, three-necked flask, equipped with a dropping funnel, a mechanical stirrer and a thermometer-outlet combination; for reaction (2): 1-1, round-bottomed flask and reflux condenser connected to a Dean–Stark water separator.

Scale: 2.0 molar for reactions (1) and 1.0 molar for reactions (2).

Procedure (for a review see Ref. [1])

1. Acetaldehyde and propionaldehyde

The commercially available aldehydes often contain considerable amounts of the cyclic trimers, even when they have been stored in the refrigerator and even when the supplier has indicated a purity of 99%! Redistillation is therefore absolutely necessary. To prevent trimerization by traces of acid adhering to the glass of the condenser and receiver, these parts of the distillation apparatus must be rinsed with a dilute solution of diethylamine or triethylamine in acetone and subsequently be blown dry prior to the distillation. The aldehydes are collected in a strongly cooled ($< -10\,^{\circ}$C) receiver and must be used within a few hours after the redistillation. Acetaldehyde can also be obtained by acid-catalyzed depolymerization of the trimer. Concentrated sulfuric acid (~ 2 ml) is added with vigorous swirling to 100–300 ml of the cyclic trimer (paraldehyde), after which the acetaldehyde is slowly distilled off through a 40-cm Vigreux column (in order to prevent vigorous decomposition, a residue of at least 15 ml should be left behind in the distillation flask).

2. Condensation of acetaldehyde and propionaldehyde with t-butylamine

t-Butylamine (2.3 mol, 167.9 g) is placed in the flask. After cooling to 15 °C, freshly distilled acetaldehyde (2.0 mol, 88.0 g) or propionaldehyde (2.0 mol, 116.0 g) is added dropwise or portionwise over 15 min, while allowing the temperature to rise to ca. 35 °C. The solution (turbid in the case of propionaldehyde) is kept for an additional 15 min at this temperature, then anhydrous potassium carbonate (40 g) is added over a few min with vigorous stirring. After 15 min (at 30–35 °C), the layers are separated (some K_2CO_3 may remain undissolved). The upper layer is transferred into the reaction flask and stirred for 15 min with 25 g of potassium carbonate, which now remains in suspension. The light-coloured liquid is then carefully decanted from the solid and transferred into a 1-l distillation flask. After cooling to room temperature, 30 g of finely machine-powdered potassium hydroxide is added. The mixture is vigorously shaken for ~ 5 min, after which the flask is connected to a 30-cm Vigreux column, condenser, and receiver cooled in a bath at − 70 °C. A tube filled with KOH pellets is placed between the water-aspirator and the receiver. The distillation apparatus is evacuated (10–20 mmHg) and the flask gradually heated to 50–60 °C. The distillate is warmed to 0 °C and then shaken for 5 min with a second portion of 25 g of machine-powdered KOH. The operation described above is repeated. Careful redistillation of the contents of the receiver through a 40-cm Widmer column gives, after a small forerun of t-butylamine, the imines $CH_3CH=N-t$-Bu, b.p. 82 °C/760 mmHg, $n_D(20)$ 1.4028, and $C_2H_5CH=N-t$-Bu, b.p. 103 °C/760 mmHg, $n_D(20)$ 1.4090, in greater than 75% yields.

The imines decompose upon standing at room temperature, but can be kept unchanged at − 20 °C for a few weeks under inert gas in well-closed bottles. It is advisable, however, to use the freshly redistilled imines for syntheses. The redistillations usually give a forerun, mainly consisting of t-butylamine, and a higher-boiling residue (compare Ref. [2]).

3. Condensation of acetaldehyde with cyclohexylamine

Freshly distilled acetaldehyde (see under 1, 96.8 g, 2.2 mol) is added dropwise to a mixture of 2.0 mol (198 g) of cyclohexylamine and 200 ml of Et_2O. The temperature is initially kept between 0 and 10 °C, but is gradually allowed to rise to ~ 25 °C in the last stage of the addition. Anhydrous potassium carbonate (60 g) is then added and the mixture is vigorously stirred for 15 min. The solid is then filtered off on a sintered glass funnel and rinsed well with Et_2O. The filtrate is cooled to 0 °C and subsequently stirred for 15 min with 50 g of freshly machine-powdered potassium hydroxide, keeping the temperature below 10 °C. After filtration and rinsing of the brown slurry with Et_2O, the solution is concentrated under reduced pressure and the remaining liquid carefully distilled through a 40-cm Widmer column to give $CH_3CH=N-c$-C_6H_{11}, b.p. 44 °C/12 mmHg, $n_D(20)$ 1.4568, in greater than 80% yield.

4. Condensation of cyclopentanone and cyclohexanone with cyclohexylamine

Cyclohexylamine (1.2 mol, 118.8 g), cyclopentanone (1.0 mol, 84.0 g), or cyclohexanone (1.0 mol, 98.0 g) and benzene (125 ml) are placed in the flask. The mixture is heated under reflux for 6 h. Almost the stoichiometrical amount of water is collected. Most of the benzene is then removed under reduced pressure (rotary evaporator) after which the products are isolated by distillation through a 30-cm Vigreus column. The imines, b.p. 108 C/12 mmHg, $n_D(20)$ 1.4946 and b.p. 125 °C/12 mmHg (solidifying at 20 °C), respectively are obtained in ~ 90% yields.

Literature

1. Layer RW (1963) Chem Rev 63:489
2. Tiollais R (1947) Bull Soc Chim France 716

4 Organic Syntheses Procedures

Wittig G, Hesse A (1970) Org Synth 50:66:

$$CH_3CH{=}O + c\text{-}C_6H_{11}NH_2 \xrightarrow{Na_2SO_4} CH_3CH{=}N{-}c\text{-}C_6H_{11} \xrightarrow[2.\,Ph_2C=O]{1.\,LDA}$$

$$Ph_2C(OH)CH_2CH{=}N{-}c\text{-}C_6H_{11} \xrightarrow{H^+,H_2O,100°C} Ph_2C{=}CHCH{=}O$$

Evans DA, Domeier LA (1974) Org Synth 54:93:

$$PhCOCH_3 + CH_3NH_2 \xrightarrow{TiCl_4} PhC(CH_3){=}N{-}CH_3$$

$$\xrightarrow[2.\,Br(CH_2)_3Cl]{1.\,LDA,\,THF} \underset{\underset{CH_2(CH_2)_3Cl}{\big|}}{Ph{-}C{=}NCH_3} \xrightarrow{reflux}$$

Perkins M, Beam Jr CF, Dyer MCD, Hauser CR (1976) Org Synth 55:39:

$$p\text{-}Cl{-}PhC(CH_3){=}NOH \xrightarrow[0°C]{2\,BuLi,\,THF} \underset{\underset{CH_2Li}{\big|}}{p\text{-}Cl{-}Ph{-}C{=}NOLi}$$

$$\xrightarrow[2.\,H^+,H_2O,reflux]{1.\,p\text{-}CH_3O{-}C_6H_4{-}COOCH_3}$$

Chapter VIII
Metallation of Nitriles and Isonitriles

1 Introduction

As in other fields of polar organometallic chemistry, the (commercial) availability of strongly basic reagents and the increased knowledge about their interaction with organic compounds have considerably stimulated the chemistry of nitriles and isonitriles.

Before 1960, only sodamide and potassium amide were used for the α-metallation of nitriles ($>$CH$-$C$\equiv$N) and successful couplings with electrophiles were restricted to alkylations [1a]. The extension of the number of base-solvent systems allowed a clean α-metallation of a variety of nitriles and the subsequent successful reaction with other nucleophiles [1b]. The use of α-metallated dialkyl-aminonitriles $R_2NCH(R')C\equiv N$ and protected cyanohydrins $R'CH(C\equiv N)OR$ ($R = SiMe_3$ or $CH(CH_3)OEt$) as acyl-anion equivalents in organic synthesis has been reviewed by Albright [2].

α-Metallated isonitriles were introduced by Schöllkopf in 1968. The versatile chemistry based upon these intermediates is the subject of a number of reviews [3–7]. Several useful synthetic applications of the tosyloxy derivatives $TsOCH_2N=C$ and its anionic intermediates have been found by van Leusen and co-workers [8, 9].

The procedures in Sect. 4 give some representative examples of metallation of nitriles and isonitriles and reactions of the anionic intermediates with alkylating agents, epoxides, aldehydes, and ketones. Syntheses involving the generation of anionic intermediates (mostly in small concentrations) and their immediate further reaction with an electrophile present in the medium during this generation fall beyond the scope of this book.

Literature

1. a. Cope AC, Holmes HL, House HO (1957) Org Reactions 9:107, John Wiley; b. Arseniyadis S, Kyler KS, Watt DS (1984) Org Reactions 31:1, John Wiley
2. Albright JD (1983) Tetrahedron 39:3207
3. Schöllkopf U (1970) Angew Chemie 82: 795; (1970) Int. ed. (Engl.) 9: 763
4. Hoppe D (1974) Angew Chemie 86:878; (1974) Int ed (Engl) 13:789
5. Schöllkopf U (1977) Angew Chemie 89:351; (1977) Int ed (Engl) 16:339
6. Schöllkopf U (1979) New Synthetic Methods, Vol 6, Verlag Chemie, Weinheim p 99
7. Schöllkopf U (1979) Pure Appl Chem 51:1347
8. van Leusen AM (1980) Heterocycl Chem (5), S 111
9. van Leusen AM (1987) In: Zwanenburg B, Klunder AJH (eds) Perspectives in the organic chemistry of sulfur, Elsevier, p 119

2 α-Metallation of Nitriles

Complete and fast metallation of nitriles $RR'CHC\equiv N$ (R and R'=H, alkyl or aryl) is effected by alkali amides in liquid ammonia (see, for instance, Ref. [1]). Also N,N-dialkylaminonitriles $R_2NCH(R')C\equiv N$ (R' = H, alkyl or aryl) react smoothly [2]. In organic solvents, e.g., toluene, addition of the alkali amide across the $C\equiv N$ bond (formation of $H-C-C(NH_2)=NM$) may seriously compete with the deprotonation [3]. Attempts to metallate allyl cyanides, $H_2C=CHCH_2C\equiv N$, with alkali amides in liquid ammonia gave mainly resinous material [2].

Interaction between nitriles and alkyllithium generally results both in addition across the $C\equiv N$ bond and in α-metallation [4]. A clean metallation can be achieved only in the cases of $CH_3C\equiv N$ and $PhCH_2C\equiv N$, using butyllithium in THF-hexane mixtures [1]. The latter nitrile can even be dilithiated under more forcing conditions [5].

The most generally applicable reagents for the α-metallation of nitriles in organic solvents are lithium dialkylamides, in particular LDA (e.g., Refs. [6] and [7]).

Literature

1. Kaiser EM, Hauser CR (1968) J Org Chem 33:3402
2. Heus-Kloos YA, Brandsma L (unpublished)
3. Ziegler K, Ohlinger H (1932) Justus Liebigs Ann Chem 495:84
4. Larcheveque M, Mulot P, Cuvigny T (1973) J Organomet Chem 57:C33
5. Kaiser EM, Solter LE, Schwarz RA, Beard RD, Hauser CR (1971) J Am Chem Soc 93:4237
6. Ahlbrecht H, Pfaff K (1978) Synthesis 897
7. Deuchert K, Hertenstein U, Hünig S (1973) Synthesis 777

3 Metallation of Isonitriles

The acidifying action of the isocyano group has been ascribed to dipole stabilization (compare Refs. [1] and [2]). Reaction of methyl isocyanide, $CH_3N=C$, with sodamide in liquid ammonia, followed by addition of a higher alkyl bromide afforded the homologue $RCH_2N=C$ in a fair yield [3]. It may be concluded from this result that sodamide causes 'ionization' to an appreciable extent at least. A very smooth and clean lithiation of $CH_3N=C$ is effected by butyllithium in a mixture of THF and hexane at temperatures between -90 and $-60\,°C$. In view of the instability of $LiCH_2N=C$ in this solvent system, derivatization reactions have to be carried out at temperatures as low as possible (above $-30\,°C$ the decomposition of $LiCH_2N=C$ becomes serious [3]). Conditions for the complete metallation of a number of isocyanides and for their partial metallation in the presence of a reaction partner have been listed by Hoppe in his review [4].

Literature

1. Beak P, Farney R (1973) J Am Chem Soc 95:4771
2. Walborsky HM, Periasamy MP (1974) J Am Chem Soc 96:3711
3. Brandsma L (unpublished)
4. Hoppe D (1974) Angew Chemie 86:878; (1974) Angew Chemie Int ed (Engl) 13:789

4 Experiments

All temperatures are internal, unless indicated otherwise.

All reactions are carried out in an atmosphere of inert gas, except those in liquid ammonia.

For general instructions concerning handling organolithium reagents, drying solvents, etc., see Vol. 1.

4.1 Metallation of Nitriles with Alkali Amides in Liquid Ammonia

$$RCH_2C \equiv N + MNH_2 \xrightarrow[-33\,°C]{\text{liq. } NH_3} RCH(M)C \equiv N + NH_3$$

$$R = H, \text{ alkyl, Ph, } Et_2N; \ M = Li, Na$$

Apparatus: 1-l round-bottomed, three-necked flask equipped with a dropping funnel, a mechanical stirrer, and a vent (rubber stopper with a hole of ca. 5 mm diameter).

Scale: 0.3 molar

Procedure

The nitrile (0.3 mol, Note 1) is added over 10 min to a well-stirred suspension of 0.3 mol of lithium amide or sodamide in 400 ml of liquid ammonia prepared as described in Vol. 1, Chap. I (Note 2). The resulting greyish solution can be immediately used for reactions with alkylating reagents. Reaction of the protected cyanohydrin $CH_3CH(C \equiv N)OCH(CH_3)OC_2H_5$ with $LiNH_2$ or $NaNH_2$ does not result in the formation of the desired intermediates.

Notes

1. For dying nitriles see page 225.
2. A small excess (< 0.3 g of Li or < 0.6 g of Na) is used to compensate for losses due to the reaction with the Fe^{III}-salt and with water dissolved in the ammonia (the water content should be not higher than 0.1%).

4.2 Lithiation of Nitriles and Isonitriles in a Mixture of THF and Hexane

$$RCH_2C\equiv N + BuLi \xrightarrow[-80\to-50\,°C]{\text{THF-hexane}} RCH(Li)C\equiv N + C_4H_{10}$$

$$(R = H \text{ or } Ph)$$

$$CH_3N=C + BuLi \xrightarrow[-80\to-65\,°C]{\text{THF-hexane}} LiCH_2N=C + C_4H_{10}$$

$$RCH_2C\equiv N + LDA \xrightarrow[-70\to-40\,°C]{\text{THF-hexane}} RCH(Li)C\equiv N + HDA$$

$$(R = H, Ph, \text{ alkyl or vinyl})$$

Apparatus: p. 24, Fig. 1; 500 ml

Scale: 0.1 molar

Procedure

Metallation with BuLi

A solution of 0.105 mol of BuLi in 70 ml of hexane is added (by syringe) to 80 ml of THF cooled to below − 40 °C. A mixture of 0.1 mol (4.1 g) of dry acetonitrile (see Exp. 1, Note 1), methyl isocyanide (4.1 g) or phenylacetonitrile (11.7 g) and 20 ml of THF is then added over 5 min, while keeping the temperature between − 80 and − 70 °C (occasional cooling in liquid nitrogen). In the cases of methyl cyanide and methyl isocyanide, a white suspension is formed immediately, the addition of phenyl acetonitrile gives a yellowish clear solution. After the addition, the temperature is allowed to rise to − 50 °C ($CH_3C\equiv N$ and $PhCH_2C\equiv N$) or − 65 °C ($CH_3N=C$), after which rections with electrophiles can be carried out. In view of the limited stability (development of a brown colour above − 50 °C), the suspension of $LiCH_2N=C$ should be used without delay.

Metallation with LDA

A mixture of 0.105 mol of the nitrile and 20 ml of THF is added over 10 min to a solution of 0.1 mol of LDA (see Vol. 1, Chap. I) in ∼ 70 ml of THF and 68 ml of hexane, while keeping the temperature between − 70 and − 60 °C. After an additional 5 min, the cooling bath is removed and the temperature allowed to rise to − 40 °C.

Aminonitriles, e.g., $Et_2NCH_2C\equiv N$, and protected cyanohydrins, e.g., $CH_3CH(C\equiv N)OCH(CH_3)—OC_2H_5$, can be lithiated in a similar way.

4.3 Alkylation of Metallated Nitriles in Liquid Ammonia and in Organic Solvents

$$RCH(M)C\equiv N + R'X \longrightarrow R'—CH(R)—C\equiv N + MX \quad M = Li \text{ or } Na,$$
$$X = \text{halogen}$$

Apparatus: 1-l round-bottomed, three-necked flask equipped with a dropping funnel, a mechanical stirrer and a vent (rubber stopper with a hole of ~ 5 mm diameter) for alkylations in liquid NH_3; p. 24, Fig. 1; 500 ml, for alkylations in organic solvents.

Scale: 0.3 molar for liq. NH_3, 0.1 molar for organic solvents.

Introduction

Nitrile anions, $\diagup{C}—C\equiv N$, are more reactive towards alkyl halides and epoxides (β-hydroxy-alkylation) than anions derived from ketones, esters, and N,N-dialkylcarboxamides [1–8]. Their solutions in liquid ammonia react vigorously with primary alkyl iodides and bromides and even with the (primary) chlorides a smooth coupling takes place at the boiling point ($- 33\,°C$) of this solvent. In mixtures of THF and hexane the alkylations with primary alkyl bromides proceed readily between $- 50$ and $0\,°C$ making the use of HMPT as co-solvent (compare Refs. [2] and [3]) superfluous. Of the secondary alkyl bromides, isopropyl bromide and cyclopentyl bromide react smoothly. The results of alkylations with cyclohexyl bromide depend strongly upon the substituents in the nitrile. Whereas $LiCH_2C\equiv N$ or $NaCH_2C\equiv N$ in liquid ammonia give cyclohexyl acetonitrile in less than 5% yield [1] (the predominant reaction being formation of cyclohexene), $PhCH(Na)C\equiv N$ (prepared in liquid ammonia) reacts with cyclohexyl bromide in toluene at elevated temperatures to give $PhCH(c\text{-}C_6H_{11})C\equiv N$ in a high yield [4]. Cyclohexylation of the lithiated aminonitrile $PhC(Li)(NMe_2)C\equiv N$ in a THF-hexane mixture has been reported to proceed successfully [5].

A serious problem in the alkylation of nitriles is the occurrence of subsequent introduction of a second substituent:

$$RCH_2C\equiv N \xrightarrow{\text{base}} R\underset{\mid}{C}H—C\equiv N \xrightarrow{R'\text{hal}} RR'CH—C\equiv N$$

$$\xrightarrow[\text{proton transfer}]{\mid} RR'\underset{}{C}—C\equiv N \xrightarrow{R'\text{hal}} RR'CR'—C\equiv N$$

In many cases this reaction strongly reduces the yield of the mono-alkylation product [2, 3, 6, 7, 9]. The order of addition seems to have little influence upon the ratio of mono- and dialkylation product [1]: it depends strongly upon the nature of the alkylating agent R hal and and upon the substituents R in the nitrile.

Disubstitution is not observed in reactions of metallated nitriles with epoxides and with aldehydes and ketones.

Procedure

Alkylation in liquid ammonia

The halide (0.32 mol) is added dropwise over 10 min to the vigorously stirred solution of 0.3 mol of the metallated nitrile in ~ 400 ml of liquid ammonia (Exp. 1). Primary alkyl bromides react vigorously and a few minutes after the addition the reaction is finished, cyclopentyl bromide reacts somewhat less vigorously. After an additional half hour, the ammonia is evaporated by placing the flask in a water bath at 30–40 °C (in the case of strong foaming the contact with the water bath should be disrupted, or small amounts of Et_2O should be occasionally added). After addition of 200 ml of water to the remaining slurry, the nitrile is isolated by extraction with Et_2O, drying of the extracts over $MgSO_4$ and careful distillation after removal of the solvent under reduced pressure. The following products have been obtained:

$C_6H_{13}CH_2C{\equiv}N$, b.p. 85 °C/15 mmHg, $n_D(20)$ 1.4208, in 76% yield from $LiCH_2C{\equiv}N$ and $C_6H_{13}Br$. The small (~ 5 g) high-boiling residue is mainly $(C_6H_{13})_2CHC{\equiv}N$;

$CH_3CH(C_4H_9)$ $C{\equiv}N$, b.p. 58 °C/15 mmHg, $n_D(21)$ 1.4093, in 60% yield from $CH_3CH(Na)C{\equiv}N$ and C_4H_9Br. The higher-boiling fraction (b.p. ~ 100 °C/15 mmHg) consists mainly of $CH_3C(C_4H_9)_2C{\equiv}N$;

$Et_2NCH(C{\equiv}N)C_4H_9$, b.p. 105 °C/15 mmHg, $n_D(20)$ 1.4360, in 71% yield from $Et_2NCH(Na)C{\equiv}N$ and C_4H_9Br (for the preparation of $Et_2NCH_2C{\equiv}N$ see Org Synth, Coll Vol III, 275 (1955));

Cyclopentyl-$CH_2C{\equiv}N$, b.p. 70 °C/12 mmHg, $n_D(22)$ 1.4473, in 65% yield from $LiCH_2C{\equiv}N$ and cyclopentyl bromide.

The reactions of $LiCH_2C{\equiv}N$ with $PhCH_2Br$ and $BrCH_2CH(OC_2H_5)_2$ gave the mono-alkylation products in ~ 30% yields. There were considerable residues consisting mainly of the dialkylation products.

Alkylation in THF-hexane mixtures

The halogen compound (0.11 mol) is added over 5 min to a solution or suspension ($LiCH_2C{\equiv}N$) of the lithiated nitrile in THF and hexane (Exp. 2) while keeping the temperature between − 60 and − 30 °C. After the addition, the cooling bath is removed and the temperature allowed to rise to 0 °C. Water (100 ml) is then added with vigorous stirring and the aqueous layer is extracted four times with Et_2O. The unwashed extracts are dried over $MgSO_4(K_2CO_3$ in the case of protected cyanohydrins) and subsequently concentrated under reduced pressure. Careful distillation (separation of the mono- and dialkylated product) gives the desired product. The following compounds have been prepared in this way:

$PhCH_2CH_2C{\equiv}N$, b.p. ~ 75 °C/0.03 mmHg, $n_D(20)$ 1.5220, in 55% yield from $LiCH_2C{\equiv}N$ and benzyl bromide (with the *chloride* impure $PhCH_2CH_2C{\equiv}N$ was obtained in a low yield, the main products were stilbene and $(PhCH_2)_2CHC{\equiv}N$);

$CH_3CH(C_6H_{13})C{\equiv}N$, b.p. 90 °C/15 mmHg, $n_D(21)$ 1.4210, in 55% yield, from $CH_3CH(Li)C{\equiv}N$ and $C_6H_{13}Br$;

Cl(CH$_2$)$_3$CH$_2$C≡N, b.p. 97 °C/15 mmHg, n$_D$(23) 1.4475, in 50% yield from
 LiCH$_2$C≡N and Br(CH$_2$)$_3$Cl;

CH$_3$C(C$_4$H$_9$)(C≡N)OCH(CH$_3$)OC$_2$H$_5$, b.p. 105 °C/15 mmHg, n$_D$(20) 1.4266, in
 70% yield from CH$_3$C(Li(C≡N)OCH(CH$_3$)OC$_2$H$_5$ and C$_4$H$_9$Br (for the
 preparation of CH$_3$CH(OH)C≡N and the protected cyanohydrin, CH$_3$CH
 (C≡N)OCH(CH$_3$)OC$_2$H$_5$ see Exp. 7);

H$_2$C=CHCH(C$_4$H$_9$)C≡N, b.p. 70 °C/12 mmHg, n$_D$(23) 1.4289, in 74% yield, from
 H$_2$C≡CHCH(Li)C≡N and C$_4$H$_9$I.

Literature

1. Brandsma L (unpublished)
2. Larcheveque M, Mulot P, Cuvigny T (1973) J Organomet Chem 57:C33
3. Cuvigny T, Normant H (1971) Organometal Chem Synth 1:237
4. Hancock EM, Cope AC (1955) Org Synth Coll III:219
5. Ahlbrecht H, Raab W, Vonderheid C (1979) Synthesis 127
6. Watt DS (1974) Tetrahedron Lett 707
7. Makosza M (1968) Tetrahedron 24:175
8. MacPhee JA, Dubois JE (1980) Tetrahedron 36:775
9. Brenner S, Bovete M (1974) Tetrahedron Lett 1377; (1975) Tetrahedron 31:153

4.4 Reaction of Metallated Nitriles with Aldehydes and Ketones

$$RCH(Li)C≡N + R'R''C=O \xrightarrow[\text{THF-hexane}]{\text{liq. NH}_3 \text{ or}} R—CH—C≡N$$

$$R'—C(R'')—OH$$

(after hydrolysis)

Apparatus: 1-1 round-bottomed, three-necked flask and p. 24, Fig. 1, 500 ml, for
reactions in ammonia and in THF-hexane, respectively.

Scale: 0.2 molar for couplings in liquid ammonia, 0.1 molar for reactions in THF-
hexane mixtures.

Introduction

Successful condensations of metallated nitriles with carbonyl compounds in liquid
ammonia are limited in number. Kaiser and Hauser obtained good results in
reactions of LiCH$_2$C≡N and NaCH$_2$C≡N with benzophenone [1]. We obtained
ca. 60 and 90% yields in the couplings of LiCH$_2$C≡N with cyclohexanone and
methyl t-butylketone, respectively, but α-lithiopropionitrile gave poor results,
presumably due to extensive enolization of the ketones [2]. As appears from
literature [1, 3, 4], lithioacetonitrile as well as substituted lithiated nitriles,
R'R''C(Li)C≡N in organic solvents, usually THF, react with a variety of carbonyl
compounds to give the corresponding carbinols in high yields. French investigators

[3] have carried out a number of these condensations in the presence of HMPT. Although (as concluded from the yields mentioned) this solvent does not seem to promote the enolization of the carbonyl compounds, it may give rise to difficulties during the isolation (extraction procedure!) of carbinols with relatively low molecular weights which have a good water-solubility. In our opinion, HMPT can be omitted in all cases.

In this experiment we describe inter alia a special procedure of working up, particularly suitable for compounds that have a good water-solubility. After reaction with the carbonyl compound a mixture of THF and water is added with vigorous stirring. The amount of water used is about 3 to 4 times that stoichiometrically required. This addition, which results in the formation of a fine suspension of $Li(OH) \cdot nH_2O$, should not be carried out at temperatures below $-10\,°C$, since then part of the water may crystallize out.

Procedure

Reaction in liquid ammonia

The ketone (0.2 mol) is added over 5 to 10 min to a vigorously stirred solution of $LiCH_2C\equiv N$ (Exp. 1) in ~ 300 ml of liquid ammonia. The ammonia is then removed by placing the flask in a water bath at $40\,°C$, or is allowed to evaporate. The solid residue is hydrolyzed with 200 ml of water, after which the products are isolated by extraction with Et_2O, drying the (unwashed) extracts over $MgSO_4$ and removing the solvent in vacuo. The following compounds have been prepared:

$(CH_2)_5C(OH)CH_2C\equiv N$; b.p. $136\,°C/15$ mmHg, $n_D(20)$ 1.4820, in 57% yield from $LiCH_2C\equiv N$ and cyclohexanone;

$t\text{-}BuC(CH_3)(OH)CH_2C\equiv N$, b.p. $118\,°C/15$ mmHg, $n_D(20)$ 1.4555, in 89% yield from $LiCH_2C\equiv N$ and methyl t-butylketone.

Reaction in THF-hexane

The freshly distilled aldehyde (0.12 mol) or cyclohexanone (0.1 mol) is added over a few s to a vigorously stirred solution or suspension ($LiCH_2C\equiv N$) of the lithiated nitrile (Exp. 2) in a mixture of THF and hexane, cooled at about $-80\,°C$. The cooling bath is removed and the temperature allowed to rise to $-20\,°C$. The product from the reaction of cyclohexanone is worked up in the usual way: addition of water (100 ml), extraction with Et_2O, drying over $MgSO_4$, and distillation.

In the other cases a mixture of 5 g of water and 30 ml of THF is added with vigorous stirring. The suspension is cautiously poured on a 1-cm layer of $MgSO_4$ or K_2CO_3 in a G-2 sintered-glass funnel (~ 5 cm diameter) with gentle sucking. The drying agent is rinsed 5 times with warm ($40\,°C$) THF, The clear filtrate is concentrated in vacuo and the remaining liquid distilled. The following compounds have been prepared:

$(CH_2)_5C(OH)CH_2C\equiv N$, b.p. $130\,°C/12$ mmHg, $n_D(20)$ 1.4820, in 85% yield from cyclohexanone and $LiCH_2C\equiv N$;

$CH_3CH(OH)CH_2C{\equiv}N$, b.p. $100\,°C/15\,mmHg$, $n_D(20)$ 1.4310, in 82% yield from $LiCH_2C{\equiv}N$ and acetaldehyde;

t-$BuCH(OH)CH(CH_3)C{\equiv}N$, b.p. $\sim 80\,°C/0.7\,mmHg$, $n_D(21)$ 1.4465, in 86% yield from $CH_3CH(Li)C{\equiv}N$ and pivaldehyde.

Literature

1. Kaiser EM, Hauser CR (1968) J Org Chem 33:3402
2. Rikers R, Brandsma L (unpublished)
3. Cuvigny T, Hullot P, Larcheveque M (1973) J Organometal Chem 57:C36
4. Albright JD (1983) Tetrahedron 39:3207

4.5 Reaction of Lithioacetonitrile with Epoxides

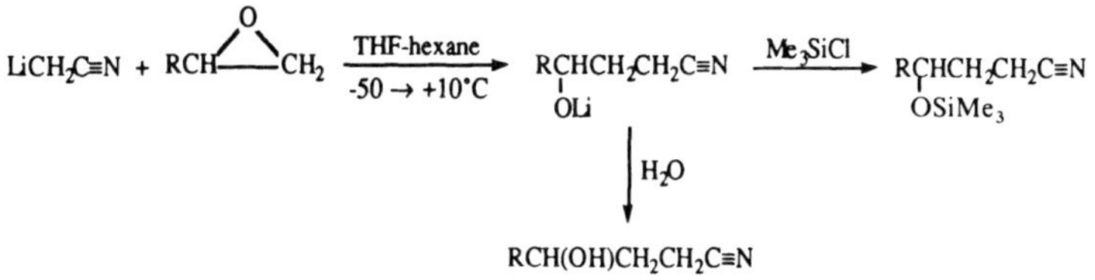

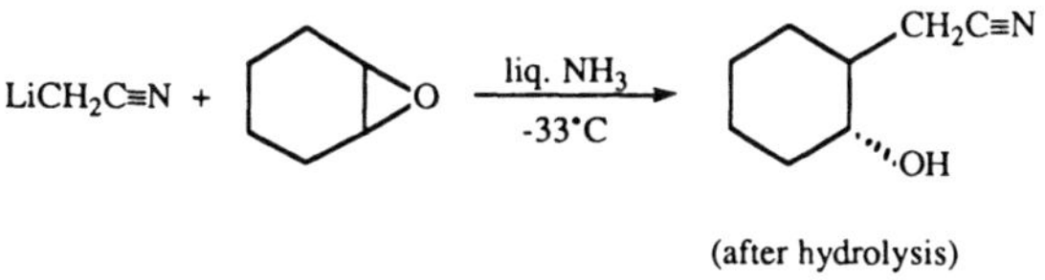

(after hydrolysis)

Apparatus: p. 24, Fig. 1; 500 ml and 1 l three-necked, round-bottomed flask, equipped with a dropping funnel, a mechanical stirrer, and a gas outlet.

Scale: 0.1 molar for reactions in THF and 0.2 molar for reactions in liquid NH_3

Procedure

Reactions in liquid ammonia

Cyclohexenoxide (0.2 mol, 19.6 g) is added over a few s to a solution of 0.2 mol of $LiCH_2C{\equiv}N$ in 500 ml of liquid ammonia (Exp. 1). The flask is insulated in cotton wool and the mixture is stirred for 2 h, then the ammonia is removed by placing the flask in a bath at 30–40 °C (in the case of strong foaming, the contact with the bath should be disrupted, or small amounts of Et_2O should be occasionally added). Water (100 ml) is added to the residue after which five extractions with Et_2O are carried out. The (unwashed) ethereal solutions are dried over $MgSO_4$ and subsequently concentrated in vacuo. Distillation through a short column affords the product, b.p. $\sim 105\,°C/0.4\,mmHg$, $n_D(21)$ 1.4809 in 85% yield.

Volatile epoxides such as oxirane and propylene oxide (0.3 mol, excess) are added dropwise over 15 min (as solutions in THF), while keeping the temperature between -35 and $-40\,°C$. After an additional half hour (at the b.p. of the ammonia), 0.3 mol of powdered ammonium chloride is added in 0.5 g portions over 15 min (Et_2O may be occasionally added to suppress foaming). The flask is then placed in a water bath at 30–40 °C, stirring being continued. When the volume of the reaction mixture has decreased to ~ 150 ml, 200 ml of THF is added and heating (now in a bath at 50 °C) is continued until the flow of escaping ammonia vapour has become very faint. The salt is filtered off on a sintered-glass funnel (with suction) and the flask and the salt on the filter are rinsed several times with small portions of warm (40 °C) THF. The filtrate is concentrated in vacuo and the remaining liquid distilled at $< 1\,mmHg$ pressure.

Reactions in THF-hexane mixtures

Epoxypropane (8.0 g, excess) is added over 10 min to a suspension of 0.1 mol of $LiCH_2C{\equiv}N$ (Exp. 2) in THF and hexane, while keeping the temperature between -50 and $-30\,°C$. After the addition, the cooling bath is removed and the temperature allowed to rise to 10 °C. Stirring at this temperature is continued for 15 min, then the clear solution is cooled to $-20\,°C$ and freshly distilled chlorotrimethylsilane (0.12 mol, 13.0 g) is added in one portion. The cooling bath is removed and the mixture is stirred for an additional 15 min at 40 °C. The greater part of the solvent is then removed under reduced pressure using a rotary evaporator. To the remaining viscous mixture is added 150 ml of a 1:1 mixture of Et_2O and pentane. After suction filtration (sintered-glass funnel) and rinsing of the salt with the Et_2O-pentane mixture, the filtrate is concentrated under reduced pressure and the remaining liquid distilled. The product, $N{\equiv}CCH_2CH_2CH(CH_3)OSiMe_3$, b.p. 82 °C/12 mmHg, $n_D(20)$ 1.4182, is obtained in 78% yield.

If the alcohol $N{\equiv}CCH_2CH_2CH(OH)CH_3$ is to be prepared, a mixture of 5 ml of water and 30 ml of THF is added instead of Me_3SiCl and the product is isolated as described in Exp. 4.

4.6 Reaction of Lithiomethyl Isocyanide with Hexyl Bromide, Oxirane, and Cyclohexanone (Note 1)

$$LiCH_2N{=}C + C_6H_{13}Br \xrightarrow[-80\,\rightarrow\,-30\,°C]{THF\text{-}hexane} C_6H_{13}CH_2N{=}C + LiBr$$

$$LiCH_2N{=}C + oxirane \xrightarrow[-65\,\rightarrow\,-20\,°C]{} HOCH_2CH_2CH_2N{=}C$$

(after hydrolysis)

$$LiCH_2N{=}C + cyclohexanone \xrightarrow[-80\,\rightarrow\,-60\,°C]{} (CH_2)_5C(OH)CH_2N{=}C$$

(after protonation with CH_3COOH)

Apparatus: p. 24, Fig. 1; 500 ml

Scale: 0.1 molar

Procedure

Alkylation

Hexyl bromide (0.1 mol, 16.5 g) is added over 15 min to the suspension of $LiCH_2N=$
C (Exp. 2) while keeping the temperature between -80 and $-60\,°C$. After the
addition, the temperature is allowed to rise to $-55\,°C$ and the mixture is stirred for
15 min between -50 and $-55\,°C$. The temperature is then allowed to rise over
10 min to $-30\,°C$ after which the cooling bath is removed. After an additional half
hour, 50 ml of water is added with vigorous stirring to the clear, almost colourless
(Note 2) solution. The organic layer and one ethereal extract are dried over
potassium carbonate. The liquid remaining after concentration of the organic
solution under reduced pressure is carefully distilled through a 30-cm Vigreux
column to give $C_6H_{13}CH_2N=C$, b.p. $72\,°C/12\,mmHg$, $n_D(20)$ 1.4328, in $\sim 80\%$
yield.

Reaction with oxirane

A mixture of 8 g (excess) of oxirane and 20 ml of THF is added over 15 min to the
suspension of $LiCH_2N=C$ (Exp. 2) while keeping the temperature between -65
and $-60\,°C$. The temperature is then allowed to rise over 15 min to $-50\,°C$, which
level is maintained for an additional half hour. After this period (Note 2) the cooling
bath is removed and the temperature allowed to rise to $-20\,°C$. A mixture of 10 ml
of water and 40 ml of THF is then added with vigorous stirring. The solution is
decanted from the aqueous slurry, which is subsequently extracted five times with
Et_2O (vigorous shaking and decanting). The combined organic solutions are dried
over $MgSO_4$ and subsequently concentrated in vacuo. Distillation of the remaining
liquid gives the alcohol, b.p. $\sim 65\,°C/0.5\,mmHg$, $n_D(20)$ 1.4366, in 80% yield.

Reaction with cyclohexanone

Cyclohexanone (0.1 mol, 9.8 g) diluted with 20 ml of THF, is added over 5 min to the
suspension of $LiCH_2N=C$ (Exp. 2), while maintaining the temperature between
-80 and $-75\,°C$. After an additional 5 min, the cooling bath is removed and the
temperature allowed to rise to $-60\,°C$. A mixture of 0.1 mol (6.0 g) of acetic acid
(note 3) and 20 ml of THF is then added over 5 min with vigorous stirring and
cooling between -60 and $-70\,°C$. After an additional 10 min (at $-70\,°C$), 100 ml of
dry Et_2O is added and the mixture is warmed to $+10\,°C$. The solution is filtered
(with suction) through a 1-cm thick layer of anhydrous potassium carbonate on a
sintered-glass funnel (~ 5 cm diameter) and the solid on the filter is rinsed well with
Et_2O. The clear filtrate is concentrated under reduced pressure and the remaining
viscous liquid distilled through a short column to give the carbinol, b.p.
$\sim 70\,°C/0.1\,mmHg$, $n_D(20)$ 1.4812, in 82% yield.

Notes

1. In order to avoid complaints about the stench, all operations with isonitriles should be carried out in a well-ventilated hood. The glassware used for the syntheses can be cleaned with a mixture of acetone and dilute (4N) hydrochloric acid.
2. If the temperature is allowed to rise too fast, the reaction mixture turns brown, due to (unknown) side-reactions, and yields of the products are considerably lower.
3. If water is added cyclization to an oxazoline may occur.

4.7 Conversion of Acetaldehyde into the Cyanohydrine and Protection of the OH Group of the Cyanohydrine with Ethyl Vinyl Ether

$$CH_3CH{=}O + KC{\equiv}N + CH_3COOH \xrightarrow[-10 \to +20\,°C]{Et_2O}$$

$$CH_3CH(OH)C{\equiv}N + CH_3COOK$$

$$CH_3CH(OH)C{\equiv}N + H_2C{=}CHOC_2H_5 \xrightarrow[acid, -5-0\,°C]{p\text{-toluene sulf.}}$$

$$CH_3CH(C{\equiv}N)OCH(CH_3)OC_2H_5$$

Apparatus: 3-l and 1-l round-bottomed, three-necked flask, equipped with a dropping funnel, an efficient mechanical stirrer, and a thermometer combined with an outlet.

Scale: 1.0 molar

Procedure

The 3-l flask is charged with 1.5 mol (100 g) of finely powdered and dry potassium cyanide and 2 l of dry Et_2O. Freshly distilled acetaldehyde (1.0 mol, 44 g) is added in one portion at $-5\,°C$. Subsequently, 1.3 mol (78 g) of anhydrous acetic acid is added over 15 min. The mixture is stirred for 30 min at -5 to $0\,°C$, then the cooling bath (dry ice-acetone) is removed and the temperature allowed to rise to $20\,°C$. Stirring at $20–25\,°C$ is continued for 2.5 h. The very thick suspension is then filtered (with suction) on a sintered-glass funnel (diameter 15–20 cm) and the salt rinsed several times with dry Et_2O (after each rinsing suction is interrupted and the salt is pressed). In view of the presence of hydrogen cyanide it is essential to carry out all operations in a well-ventilated hood! The clear solution is concentrated under reduced pressure, leaving the almost pure cyanohydrine, $n_D(21)$ 1.4071, in $\sim 90\%$ yield.

Freshly distilled ethyl vinyl ether (2.0 mol, 140 g, excess!) is placed in the 1-l flask. p-Toluene sulfonic acid (200 mg; it is not necessary to use the *anhydrous* acid) is added at $-10\,°C$ with efficient stirring. The undistilled cyanohydrine is added dropwise over 30 min, while the temperature is kept between -5 and $0\,°C$. In order to control whether the reaction proceeds, ca. 5 ml of the cyanohydrine may be added in one portion at $-5\,°C$ (the contact with the cooling bath is temporarily disrupted): this addition should result in a fast rise of the temperature by several degrees C. When all cyanohydrine has been added, a second portion of 200 mg of p-toluenesulfonic acid is introduced at $0\,°C$: this should not result in a significant rise of the temperature. After an additional 15 min (at $0\,°C$), the almost colourless solution is cooled to $-5\,°C$ and a solution of 5 g of potassium carbonate in 15 ml of water is added with vigorous stirring and cooling between 0 and $5\,°C$. After an additional stirring period of 2 min, the layers are separated. The organic layer is dried well over anhydrous potassium carbonate, after which the solvent is completely removed under reduced pressure, keeping the bath temperature below $50\,°C$. The remaining liquid, reasonably pure according to ^1H NMR (yield almost quantitative), is used as such for syntheses. Distillation is not carried out, since decomposition may occur.

Chapter IX
Generation of Lithium Halocarbenoids

1 Introduction

Lithium halocarbenoids are species in which lithium and one, two, or three halogen atoms are linked to the same carbon atom. A great deal of the chemistry of these carbenoids was developed by Köbrich and his co-workers [1] who were the first to generate these species in solution. Useful synthetic applications were found several years later by a number of groups, which could take advantage of the considerable progress in the field of polar organometallic chemistry.

Although some successful syntheses have been realized at relatively high temperatures with transient carbenoids and very reactive trapping reagents like carbonyl compounds and chlorotrimethylsilane, halocarbenoid chemistry is in general performed below $-80\,°C$ with preformed carbenoids.

The review by Siegel [2] covers most of the halocarbenoid chemistry. The procedures in this chapter exemplify the methods and techniques used in the generation of lithium halocarbenoids and in their subsequent functionalization reactions; in accordance with the scope of this volume, only carbenoids in which the carbanionic centre is sp^3-like ($\rangle$C(Li)Hal) are considered.

Literature

1. Köbrich G (1972) Angew Chemie 84:557; (1972) Angew Chemie Int ed (Engl) 11:473
2. Siegel W (1982) Lithium Halocarbenoids, Carbanions of High Synthetic Versability, Topics in Current Chemistry 106:55, Springer-Verlag

2 Methods of Generation of Lithium Halocarbenoids

a) Deprotonation

The acidifying influence exerted by halogen atoms in haloforms, HCX_3, and 1,1-dihaloalkanes, $RCHX_2$, is stronger than that of many heterosubstituents. For chloroform and bromoform, pK values of 16 and 9, respectively, have been determined [1]. The high kinetic acidity appears also from the high yields of the carbenoids obtained by adding lithium dicyclohexylamide to a mixture of CH_2X_2 or HCX_3 (X = Cl or Br) and an enolizable aldehyde or ketone [2]. Thus, even the readily enolizable cyclopentanone, reacted efficiently in this procedure. Although in a number of cases [2–4] a such like procedure, in which the carbenoid is generated in the presence of the derivatization reagent, gives good results, the carbenoid is usually preformed at a sufficiently low temperature and functionalized in a successive operation.

The deprotonation of the 1,1-dihaloalkane, $RCHX_2$, or haloform, HCX_3, is generally carried out with a very strongly basic organolithium reagent. Butyllithium is not suitable for the generation of $RC(Li)Br_2$ and $LiCBr_3$ (or analogous iodine compounds), since it attacks also on bromine. Using LDA the carbenoids $LiCHCl_2$, $LiCHBr_2$, $LiCHI_2$, $RC(Li)Br_2$, ($R = $ alkyl or aryl), $LiCCl_3$, and $LiCBr_3$ can be generated almost quantitatively at temperatures between -100 and $-110\,^\circ C$ in mixtures of THF and hexane [5, 6]. The methylene protons in CH_2Cl_2 and the methine proton in $RCHCl_2$ ($R = $ alkyl) are kinetically less acidic, but in these cases a quick and quantitative lithiation can be achieved with butyllithium and butyllithium·TMEDA, respectively, in a THF-hexane mixture. Monohalogen carbenoids, $RCH(Li)X$ ($R = $ alkyl) cannot be prepared by deprotonation reactions. Attempts to generate the carbenoids $RCH(Li)X$ with $R = $ phenyl or allyl and $X = Cl$ or Br from RCH_2X and LDA led to the formation of $RCH(X)—CH_2R$ as a consequence of a very fast subsequent allylation or benzylation of the (transient) carbenoids [3].

b) Halogen–lithium exchange

Monohalogen carbenoids $R'R''C(Li)Hal$ (R' and $R'' = $ alkyl; $Hal = Cl$ or Br) cannot be prepared by deprotonation. A number of representatives have been successfully generated from $R'R''C(Br)$ Hal and n- or sec-BuLi, e.g., [7, 8]:

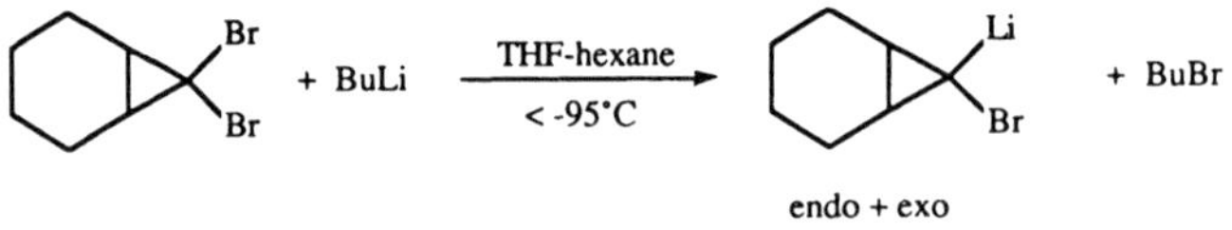

endo + exo

$$HalCH_2Br + sec\text{-BuLi} \xrightarrow[< -110\,^\circ C]{THF-Et_2O-hexane} HalCH_2Li + sec\text{-BuBr}$$

$$(Hal = Cl \text{ or } Br)$$

Literature

1. Schlosser M (1974) Polare Organometalle. Springer-Verlag, p 17
2. Taguchi H, Yamamoto H, Nozaki H (1974) J Am Chem Soc 96, 3010
3. Andringa H, Heus-Kloos YA, Brandsma L (1987) J Organometal Chem 336:C41
4. Macdonald TL, Amirthalingam Narayanan B, O'Dell DE (1981) J Org Chem 46:1504
5. Siegel H (1982) Topics in Current Chemistry 106:55. Springer-Verlag
6. Verkruijsse HD, Brandsma L (unpublished)
7. Seyferth D, Lambert Jr RL, Massol M (1975) J Organometal Chem 88:255
8. Tarhouni R, Kirschlager B, Rambaud M, Villiéras J (1984) Tetrahedron Lett 835

3 Experimental Conditions and Techniques in Carbenoid Chemistry

Most lithium halocarbenoids decompose at temperatures above -80 to $-90\,°C$. An essential condition for a successful synthesis via a (preformed) lithium halocarbenoid is that its generation proceeds at a high rate at temperatures below this range. For this reason, the more polar THF is preferred to Et_2O as a main solvent for the deprotonation or lithium-halogen exchange reactions by which the carbenoids are formed.

Because of the increase of the viscosity of THF at very low temperatures, stirring becomes less efficient. For this reason, Et_2O and/or a low-boiling petroleum ether fraction are used as co-solvent. Especially in reactions that have to be carried out below $-110\,°C$, relatively large amounts of solvents are necessary, which puts limitations on the scale of the synthesis (usually not higher than 50 mmol at these very low temperatures, unless adapted equipment is available).

The manipulations necessary to maintain the temperature within a small range of some $5\,°C$ require a considerable experimental skill and a continuous, active involvement during the performance of the reaction. During cooling with liquid nitrogen, one should prevent solidification of the solvent on the bottom of the flask: occasional contact with the liquid nitrogen during the addition of the reagents is usually sufficient for neutralizing heating effects. When the reaction does not proceed instantaneously (e.g., in the case of alkylations), the reaction mixture has to be kept for a certain additional period at a low temperature. This may be realized as follows. The greater part of the liquid nitrogen is poured out of the cooling vessel (Dewar flask, height 20 cm, diameter ~ 15 cm). The reaction flask is placed in the cooling vessel, with the bottom just above the surface of the remaining liquid nitrogen, and the upper part of the flask covered with cotton wool. The reaction mixture is stirred gently or at a moderate rate. The desired temperature range can be adjusted by varying the distance between the bottom of the flask and the level of the liquid nitrogen.

The procedure of adding the reagents needs special comments. Since at temperatures in the region of $-100\,°C$ the viscosity of the solvents has increased considerably, the added reagent is less quickly distributed over the solution in the reaction flask. The locally occurring heating effects due to very fast reactions may give rise to decomposition of the carbenoid. Addition of the reagent in a diluted state is therefore advisable.

4 Experiments

All temperatures are internal, unless indicated otherwise.
All reactions are carried out in an atmosphere of inert gas.
For general instructions concerning handling organolithium reagents, drying solvents, etc., see Vol. 1.

4.1 Lithiation of Dichloromethane

$$H_2CCl_2 + BuLi \xrightarrow[-90°C]{THF\text{-}hexane} LiCHCl_2\downarrow + C_4H_{10}$$

$$LiCHCl_2 + PhCH{=}O \xrightarrow[-95°C]{THF\text{-}hexane} PhCH(OH)CHCl_2 \quad \text{(after hydrolysis)}$$

$$LiCHCl_2 + C_8H_{17}Br \xrightarrow{THF\text{-}hexane, \ HMPT} C_8H_{17}CHCl_2 + LiBr$$
$$-90 \rightarrow -70°C$$

Apparatus: p. 24, Fig. 1; 11

Scale: 0.1 molar

Procedure (compare Ref. [1])

a) A solution of 0.1 mol of butyllithium (Note 1) in 68 ml of hexane is cooled to below − 10 °C and 70 ml of THF (Note 2) is added. Dichloromethane (0.12 mol, 10.2 g) is added over 5 min with cooling between −90 and − 100 °C and stirring at a moderate rate. After an additional 10 min (at ~ − 90 °C), the white suspension is ready for functionalization reactions.

b) Freshly distilled benzaldehyde (0.09 mol, 9.5 g) diluted with 20 ml of THF or Et$_2$O is added over 5 min with cooling between − 85 and − 100 °C and vigorous stirring. After an additional 5 min, the cooling bath is removed and 200 ml of water is added with vigorous stirring. After separation of the layers, one extraction with Et$_2$O is carried out. The organic solution is dried over MgSO$_4$ and subsequently concentrated under reduced pressure. Distillation of the remaining liquid through a very short column gives the carbinol, b.p. ~ 105 °C/1 mmHg, n$_D$(20) 1.5623, in almost quantitative yield.

c) n-Octyl bromide (0.09 mol, 17.4 g) is added over a few s with cooling between − 90 and − 100 °C. Subsequently, a mixture of 20 g of dry HMPT (see Vol. 1, Chap. I and p. 225 for the purification of HMPT) and 20 ml of THF is added dropwise over 10 min with stirring at a moderate rate, while maintaining the temperature between − 90 and − 100 °C. After stirring for an additional half hour at ~ − 80 °C, the suspended material has dissolved completely. The cooling bath is removed and the temperature allowed to rise to − 40 °C. Dilute (2 N) hydrochloric acid (150 ml) is added with vigorous stirring, after which the layers are separated. The aqueous layer is extracted once with a small amount of pentane. The organic solution is washed three times with water, dried over MgSO$_4$ and subsequently concentrated in vacuo. Distillation of the remaining liquid through a 30-cm Vigreux column gives the alkylation product, b.p. 103 °C/15 mmHg, n$_D$(22) 1.4487, in greater than 90% yield.

Notes

1. With LDA in THF and hexane (~ -90 to $-100\,°C$) also good results were obtained (compare Ref. [1], p. 765).
2. With BuLi in a mixture of Et_2O and hexane the metallation is very slow at $-95\,°C$. Above $-80\,°C$ metallation and decomposition (of the carbenoid) proceed at comparable rates.

Literature

1. Villiéras J, Pierot P, Normant JF (1977) Bull Soc Chim France 765

4.2 Lithiation of 1,1-Dichloroalkanes

$$RCHCl_2 + BuLi\ (TMEDA) \xrightarrow[-100--110°C]{THF\text{-}hexane} RC(Li)Cl_2 \downarrow + C_4H_{10}$$

$$RC(Li)Cl_2 + t\text{-}C_4H_9CH{=}O \xrightarrow[-100°C]{} RCl_2CCH(OH)t\text{-}C_4H_9$$

$$\text{(after hydrolysis)}$$

$$RC(Li)Cl_2 + D_2O \xrightarrow[-100°C]{} RC(D)Cl_2 + LiOD \quad (R{=}CH_3,\ C_8H_{17})$$

Apparatus: p. 24, Fig. 1; 11

Scale: 0.1 molar

Procedure (compare Ref. [1] in Exp. 1)

THF (70 ml) and 0.11 mol of BuLi in 73 ml of hexane are mixed as described in Exp. 1a. TMEDA (0.11 mol, 12.7 g, Note 1) is added with cooling below $-50\,°C$. The dichloroalkane (0.13 mol, diluted with 30 ml of THF) is then added dropwise over 10 min with cooling between -100 and $-110\,°C$. The rather thick white suspension is stirred for an additional 5 min at $-95\,°C$ (Note 2), then pivaldehyde (0.1 mol, 8.6 g, diluted with 30 ml of THF) or a mixture of 0.3 mol (6.0 g) of D_2O and 30 ml of THF is added dropwise with vigorous stirring and cooling between -90 and $-95\,°C$. After the addition, the cooling bath is removed and the temperature allowed to rise to $-40\,°C$ (in the case of the aldehyde) or $0\,°C$ (in the case of the deuteration). The mixture is treated with water (100 ml) after which the layers are separated and the aqueous layer is extracted once with Et_2O. The organic solution is dried over $MgSO_4$, after which the product is isolated as described in Exp. 1c. The carbinol, $CH_3CCl_2CH(OH)\text{-}t\text{-}C_4H_9$, b.p. $75\,°C/14$ mmHg, $n_D(22)$ 1.4662 (solidifying at room temperature), and the deuterated compound $C_8H_{17}C(D)Cl_2$, b.p. $100\,°C/12$ mmHg, are obtained in greater than 85% yields.

Notes

1. The metallation of CH_3CHCl_2 has also been carried out in the absence of TMEDA (addition of CH_3CHCl_2 over 5 min, additional time 15 min at $\sim -100\,°C$). The carbinol $CH_3CCl_2CH(OH)$—t-C_4H_9 was obtained in $\sim 75\%$ yield.
2. Above $-85\,°C$ a very exothermic decomposition of $CH_3C(Li)Cl_2$ occurs. We have not observed any positive influence of TMEDA upon the stability of $CH_3C(Li)Cl_2$.

4.3 Lithiation of Dibromomethane

$$H_2CBr_2 + LDA \xrightarrow[-100°C]{\text{THF-hexane}} LiCHBr_2\downarrow + HDA$$

$$LiCHBr_2 + Me_3SiCl \xrightarrow[-100°C \to -40°C]{} Me_3SiCHBr_2 + LiCl$$

$$LiCHBr_2 + C_4H_9Br \xrightarrow[-100 \to -20°C]{+ \text{HMPT}} C_4H_9CHBr_2 + LiBr$$

Apparatus: p. 24; Fig. 1; 11

Scale: 0.2 molar

Procedure

a) A solution of 0.2 mol of BuLi in 132 ml of hexane is added over a few min to a mixture of 0.2 mol (20.2 g) of diisopropylamine and 120 ml of THF, while keeping the temperature below $0\,°C$. The resulting solution is cooled to $-100\,°C$, after which 0.3 mol (52.2 g) of dibromomethane is added dropwise over 15 min with stirring at a moderate rate. After an additional 10 min (at $\sim -100\,°C$), the yellow suspension is used for further conversions.
b) Freshly distilled chlorotrimethylsilane (0.22 mol, 23.9 g) is added over 15 min, while keeping the temperature between -95 and $-105\,°C$. The suspension soon turns white and disappears almost completely after an additional period of 15 min of stirring at $\sim -90\,°C$. The cooling bath is then removed and the temperature allowed to rise to $-40\,°C$. Water (100 ml) is then added with vigorous stirring, after which 2 N hydrochloric acid is added to bring the pH of the aqueous layer at 4 to 6. The aqueous layer is extracted once with a small amount of pentane. After drying the organic solution over $MgSO_4$, the greater part of the solvent is distilled off at atmospheric pressure. Careful distillation of the remaining liquid through a 30-cm Widmer column gives $Me_3SiCHBr_2$, b.p. $54\,°C/14\,mmHg$, $n_D(21)$ 1.4992, in greater than 90% yield.

c) Butyl bromide (0.25 mol, 34.2 g) is added over 5 min to the suspension of $LiCHBr_2$ while cooling between -95 and $-100\,°C$. Subsequently, a mixture of 30 ml of purified HMPT (see Vol I, Chap. I and p. 225) and 20 ml of THF is added dropwise over 15 min with cooling between -90 and $-100\,°C$. The suspended material dissolves completely and a light-brown solution is formed. After an additional period of 45 min (at $-80\,°C$), the cooling bath is removed and the temperature allowed to rise to $-20\,°C$. Dilute (3 N) hydrochloric acid is then added below pH 7. The aqueous layer is extracted once with pentane. After washing the organic solution four times with water and drying over $MgSO_4$, the greater part of the solvent is distilled off at atmospheric pressure. Careful distillation of the remaining liquid through a 30-cm Widmer column gives $C_4H_9CHBr_2$, b.p. $65\,°C/15\,mmHg$, $n_D(24)$ 1.4942, in 62% yield. This product can be lithiated and alkylated with C_4H_9Br by a similar procedure. $C_4H_9CBr_2C_4H_9$, b.p. $115\,°C/15\,mmHg$, $n_D(23)$ 1.4888, is obtained in $\sim 75\%$ yield. Since during the lithiation the reaction mixture becomes very thick, 75 ml of Et_2O is added portionwise (at -105 to $-110\,°C$).

4.4 Lithiation of Chloroform

$$HCCl_3 + LDA \xrightarrow[-100 \text{ to } -110°C]{THF\text{-}hexane} LiCCl_3\downarrow + HDA$$

$$LiCCl_3 + C_5H_{11}Br \xrightarrow[-100 \to -70°C]{+HMPT} C_5H_{11}CCl_3 + LiBr$$

Apparatus: p. 24; Fig. 1; 11

Scale: 0.1 molar

Introduction

A general procedure for lithiation involves dropwise addition of a solution of the lithiating reagent to a mixture of the substrate and an organic solvent. Several carbenoids have been generated in this way. Addition of a 1.6 molar solution of BuLi in hexane to a mixture of chloroform and THF cooled to below $-100\,°C$ resulted in an extremely vigorous reaction and the formation of a very dark solution. Subsequent addition of benzaldehyde gave the carbinol $PhCH(OH)CCl_3$ in a low yield. This might be explained by decomposition of $LiCCl_3$ due to strong local heating effects. Seyferth et al. avoid this situation by allowing the solution of BuLi to run down the cold wall of the flask [1]. When the order of addition was reversed (i.e., dropwise addition of a mixture of $CHCl_3$ and THF to a solution of BuLi in THF and hexane), however, the carbenoid could be trapped with excellent yields, with the somewhat milder reagent LDA we obtained also good results. The procedure for the alkylation of trichloromethyllithium in the presence of a small amount of HMPT is based on that described by Villiéras et al. [2]. Secondary alkyl halides give poor results due to their low reactivity.

Procedure

A solution of 0.11 mol of BuLi in 73 ml of hexane is added over a few s (by syringe) to a mixture of 0.11 mol (11.2 g) of diisopropylamine and 70 ml of THF, while keeping the temperature below 0 °C. A mixture of chloroform (0.11 mol, 13.0 g) and 20 ml of THF is added dropwise over 10 min with efficient stirring and cooling between -100 and -110 °C. After an additional 5 min, 0.15 mol (22.6 g) of pentyl bromide is added over a few s to the white suspension. Subsequently, a mixture of 25 ml of purified HMPT (see Vol. 1, Chap. I and p. 225) and 20 ml of THF is added dropwise over 10 min with cooling between -100 and -105 °C. The suspended material gradually dissolves and a light-brown solution is formed. After stirring for an additional 1 h at -95 to -100 °C, the cooling bath is removed and the temperature allowed to rise to -70 °C. Dilute (2 N) hydrochloric acid (250 ml) is added with vigorous stirring. The aqueous layer is extracted once with pentane. The solution is washed 4 times with water. After drying the organic solution over $MgSO_4$, the solvent is removed in vacuo and the remaining liquid carefully distilled through a 30-cm Widmer column. After a fore-run of pentyl bromide, $C_5H_{11}CCl_3$, b.p. 65 °C/15 mmHg, $n_D(20)$ 1.4565, is collected in greater than 70% yield.

Literature

1. Seyferth D, Lambert Jr RL, Massol M (1975) J Organometal Chem 88:255
2. Villiéras J, Bacquet C, Normant JF (1975) Bull Soc Chim France 1797

4.5 Lithiation of Bromoform

$$LDA + HCBr_3 \xrightarrow[-110]{\text{THF-hexane}} LiCBr_3 \downarrow + HDA$$

$$LiCBr_3 + C_4H_9Br \xrightarrow[-100 \text{ to } -110°C]{\text{THF-hexane-HMPT}} C_4H_9CBr_3 + LiBr$$

Apparatus: p. 24, Fig. 1; 11

Scale: 0.1 molar

Procedure

A solution of 0.1 mol of LDA in 70 ml of THF and 66 ml of hexane is prepared as described in Exp. 4. Et_2O (75 ml) is added. A mixture of 0.1 mol (25.3 g) of bromoform and 40 ml of THF is added dropwise over 15 min with cooling between − 105 and − 110 °C. A rather thick yellow suspension is formed. After an additional 10 min (at ∼ − 105 °C) butyl bromide (0.15 mol, 20.5 g) is added over 30 s at − 105 °C. Subsequently, a mixture of 25 g of purified HMPT (see Vol. 1, Chap. I and p. 225) and 20 ml of THF is added over 10 min with cooling between − 100 and − 110 °C. The reaction mixture is stirred for an additional 45 min at − 105 °C, then for 15 min at − 95 °C, after which the cooling bath is removed. When the temperature of the dark-brown solution has reached − 50 °C, 250 ml of 2 N hydrochloric acid is added with vigorous stirring. The product, $C_4H_9CBr_3$, b.p. 95 °C/15 mmHg, $n_D(22)$ 1.5385, yield 85%, is isolated as described in Exp. 4.

4.6 Lithiation of 7,7-Dibromonorcarane

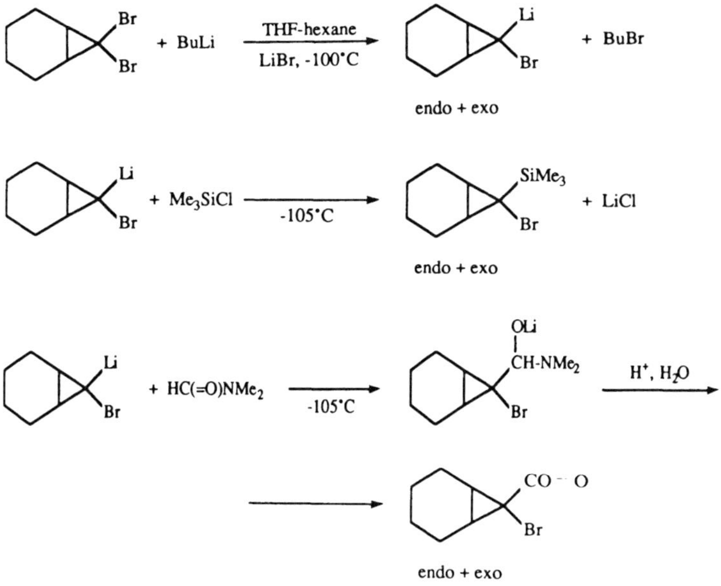

Apparatus: p. 24, Fig. 1; 11

Scale: 0.1 molar

Introduction

Seyferth et al. [1] give well-detailed descriptions for the generation of a number of lithium halocarbenoids from geminal dihalocyclopropanes. In their procedures, a solution of BuLi in hexane is added to a mixture of the dihalo compound and THF kept between − 90 and − 100 °C. According to Villiéras [2], lithium bromide stabilizes lithium halocarbenoids by forming a complex:

This might weaken the destabilizing interaction between the halogen atom and lithium in the carbenoids. An alternative explanation for the LiBr-effect is that the salt forms a mixed aggregate with BuLi which reacts somewhat less vigorously with the substrate. The local, strong heating effects which might be responsible for the decomposition of the carbenoid thus are moderated.

In Seyferth's procedures [1], which do not use lithium bromide, the BuLi solution is allowed to run down the cold glass wall before it makes contact with the solution of the substrate. The reason for applying this technique is probably also to minimize local heating effects.

In our procedure for the generation of bromolithionorcarane inversed-order addition is applied. The butyllithium is combined with (anhydrous) lithium bromide, though the difference in the results with experiments in which this salt was omitted was slight.

Procedure

a. Anhydrous lithium bromide (0.12 mol, 10.4 g; the commercially available anhydrous salt is heated for $\sim$ 30 min at 150 °C in a vacuum of <1 mmHg) is dissolved in 80 ml of THF and a solution of 0.11 mol of BuLi in 73 ml of hexane is added with cooling below 0 °C. After addition of 50 ml of Et$_2$O, the solution is cooled to between -100 and -110 °C and a mixture of 0.1 mol (25.4 g) of 7,7-dibromonorcarane and 40 ml of THF is added dropwise over 10 min, while maintaining the temperature between -100 and -110 °C. After an additional 10 min, the reaction mixture (fine suspension) is used for the reactions described below.

b. Freshly distilled chlorotrimethylsilane (0.15 mol, 16.2 g) is added dropwise over 5 min with cooling between -100 and -105 °C. After an additional half hour, the cooling bath is removed and the temperature allowed to rise to -60 °C. Water (200 ml) is then added and the aqueous layer is extracted once with pentane. The organic solution is dried over MgSO$_4$, then concentrated in vacuo and the remaining liquid carefully distilled through a 30-cm Vigreux column. A 1:1 mixture (compare Ref. 1) of *endo* and *exo* 7-bromo-7-trimethylsilyl-norcarane, b.p. 100–115 °C/14 mmHg, n$_D$(21) 1.5080, is obtained in 85% yield.

c. Dimethylformamide (0.15 mol, 11.0 g) is added over 5 min at ~ -105 °C. After an additional 15 min, the cooling bath is removed and the temperature allowed to rise to -60 °C. The solution is added over a few min to a vigorously stirred mixture of 15 g of 36% hydrochloric acid and 150 ml of water. The aqueous layer is extracted three times with pentane. The combined organic solutions are washed with a saturated solution of sodium chloride and subsequently dried over MgSO$_4$. After removal of the solvent in vacuo, the remaining liquid is distilled through a 20-cm Vigreux column to give 7-bromo-7-formylnorcarane, b.p. $\sim 65\,°C/0.3\,mmHg$, as a 1:3 mixture of the *endo*- and *exo*-isomer in 75% yield.

Literature

1. Seyferth D, Lambert Jr RL, Massol M (1975) J Organometal Chem 88:255
2. Tarhouni R, Kirschleger B, Rambaud M, Villiéras J (1984) Tetrahedron Lett 835

4.7 Lithiation of Ethyl Bromoacetate

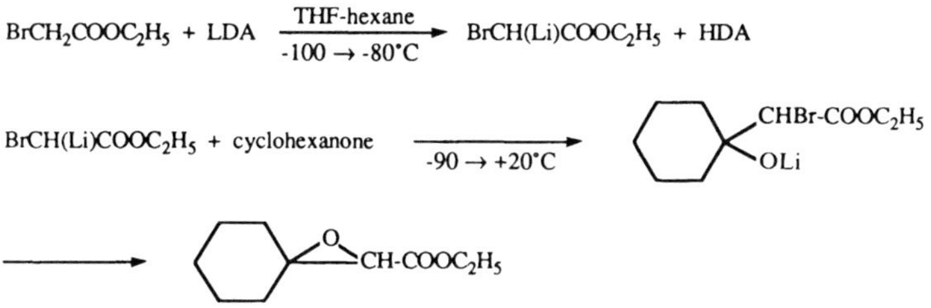

$$BrCH_2COOC_2H_5 + LDA \xrightarrow[-100 \to -80°C]{THF\text{-}hexane} BrCH(Li)COOC_2H_5 + HDA$$

$$BrCH(Li)COOC_2H_5 + cyclohexanone \xrightarrow[-90 \to +20°C]{}$$

$$\longrightarrow$$

Apparatus: p. 24; Fig. 1; 11

Scale: 0.1 molar

Introduction

Darzens glycid ester condensations of α-haloesters with aldehydes and ketones are usually carried out by treating the mixture of compounds with a relatively weakly basic reagent like ethoxide. The transient carbenoid RC(M)(Hal)COOR′ reacts under these conditions to give the alkali carbinolate, which immediately undergoes elimination of alkali halide resulting in the epoxide ester (or glycid ester). Pre-formation of the carbenoid has been effected with lithium bis(trimethylsilyl)amide at very low temperatures. The aldehyde or ketone is subsequently added after which the temperature is allowed to rise. Under these conditions, cyclization to the glycid ester takes place. This method gives good results with aliphatic aldehydes, which undergo extensive self condensation under the conventional conditions. In view of the low stability of the ester halocarbenoids it seems essential to generate these species in a fast reaction at the lowest possible temperature. LDA should therefore be preferred over the less strongly basic $LiN(SiMe_3)_2$ used by Borch [1].

Procedure

A solution of 0.105 mol of LDA in 69 ml of hexane and 80 ml of THF (see Exp. 4) is cooled to $-100\,°C$. A mixture of 0.1 mol (16.7 g) of freshly distilled ethyl bromoacetate and 20 ml of THF is added dropwise over 5 min, while keeping the temperature between − 100 and − 95 °C. The suspended material (LDA) gradually dissolves. After an additional 5 min (∼ − 80 °C), the solution is cooled again to − 95 °C and 0.11 mol (10.8 g) of cyclohexanone is added over 5 min while maintaining the temperature between − 80 and − 90 °C. After the addition, the cooling bath is removed and the temperature allowed to rise to ∼ − 40 °C. Finally, the light-brown mixture is warmed to + 20 °C, after which 100 ml of ice water is added. The organic layer and one ethereal extract are dried over $MgSO_4$ after which the glycid ester is isolated by distillation: b.p. 122 °C/12 mmHg, $n_D(21)$ 1.4600, in 72% yield.

If during the lithiation the temperature is allowed to rise above − 75 °C, a dark brown solution is formed and the product is obtained in a much lower yield.

Literature

1. Borch RF (1972) Tetrahedron Lett 376

Chapter X
Metallation of Carbonyl and Thiocarbonyl Compounds

1 Introduction

Enolates are the most important types of anionic intermediates in organic synthesis. In many of the familiar conversions, the alkali enolate is generated in the presence of another reagent with which it reacts during its generation. 'Condensations', for example, in many cases are performed by treating a mixture of an aldehyde, ketone, or carboxylic ester and the reaction partner with a basic reagent such as sodium ethoxide or sodium hydride. The conversions in this chapter are carried out with enolates that are preformed in a separate operation by reaction of the carbonyl compound with a very strong base such as LDA. These conditions allow qualitative comparisons of the reactivities of enolates and other polar organometallic intermediates.

Compared with other synthetic intermediates, enolates show a decreased reactivity. The differences in reactivity are most striking in reactions with alkylating agents [1] and epoxides [6]. The reactivities of the various types of enolates towards alkyl halides decrease in the order $C{=}C(O^-)NR_2$ (amide-enolate) $C{=}C(O^-)OR$ (ester enolate) $>C{=}CO^-$ (ketone-enolate). Metallated nitriles, imines, and S,S-acetals are, in general, much better nucleophiles than enolates in alkylations and β-hydroxyalkylations [1]. Furthermore, the alkylation of aldehyde and ketone enolates usually does not stop after the mono-functionalization [12]. The decreased reactivity of (especially) aldehyde and ketone enolates also appears in thiolations with disulfides [2]. A solution of lithiated cyclohexanone in THF does not react at $20\,°C$ with CH_3SSCH_3 [1, 2].

Enethiolates usually undergo S-alkylation with (primary) formation of S,S-, O,S-, or N,S-acetals [3, 4].

The chemistry of alkali enolates is subject of a number of extensive reviews [5–13]. The stereochemistry of directed aldol reactions with lithium enolates is discussed in the review of Mukaiyama [5].

Literature

1. Brandsma L, Verkruijsse HD (unpublished)
2. Seebach D, Teschner M (1973) Tetrahedron Lett 5113
3. Schuijl PJW, Brandsma L, Arens JF (1965) Recl Trav Chim Pays-Bas 85:1263
4. Schuijl PJW, Bos HJT, Brandsma L (1968) *ibid.* 87:123; Schuijl PJW, Brandsma L (1968) *ibid.* 87:929
5. Mukaiyama T (1982) in Org Reactions 28:203, John Wiley

6. Sturm TJ, Marolewski AE, Rezenka DS, Taylor SK (1989) J Org Chem 54:2039
7. Jackman LM, Lange BC (1977) Tetrahedron 33:2737
8. Seebach D, (1988) Angew Chemie 100: 1685; (1988) Angew Chemie Int ed (Engl) 27:1624
9. Kaiser EM, Petty JD, Knutson PLA (1977) Synthesis 509
10. Ivanov D, Vassilev G, Panayotov I (1975) Synthesis 83
11. Harris TM, Harris CM (1969) Org Reactions 17:155, John Wiley
12. d'Angelo J (1976) Tetrahedron 32:2979
13. House HO (1972) Modern synthetic reactions, 2nd ed., Chap 9, Benjamin

2 Mono-Metallation of Carbonyl and Thiocarbonyl Compounds

The very acidic (in the region of pK 10) 1,3-dicarbonyl compounds ($-CO-CH-$
$CO-$), diketones, and ketoesters are readily converted into the enolates by reaction
with alkali hydrides in polar as well as in apolar solvents [1]. Also some acidic
mono-ketones, e.g., $PhCH_2COCH_3$, react smoothly with sodium hydride [4].
Potassium hydride has been used to convert aldehydes into the enolates
[2, compare 3]. A serious drawback of this method is the handling of the
commercially available potassium hydride, which inter alia involves removal of
the protecting mineral oil. Reaction of kinetically less acidic ketones and carboxylic
esters with alkali hydrides, in general, gives rise to extensive formation of
self-condensation products [3, 6].

Although other 'sterically hindered' bases, e.g., lithium dicyclohexylamide,
lithium *bis*-(trimethylsilyl) amide, and lithium isopropyl cyclohexyl amide have been
recommended, the relatively cheap LDA gives generally good results in the
generation of enolates from a variety of carbonyl and thiocarbonyl compounds.
Treatment of non-symmetrical ketones with LDA in THF usually gives exclusively
or predominantly the least stable ('kinetic') enolates [4, 5]. Thus, 2-methylcylohexa-
none, methyl allyl ketones, methyl alkenyl ketones ($CH_3CO-C=C$), and methyl
alkynyl ketones ($CH_3CO-C\equiv C$) [6] deprotonation with LDA undergo at the
methylene and methyl groups, respectively, with very high selectivity.

The deprotonations with LDA are generally extremely fast, even at temperatures
far below 0 °C and side- or subsequent reactions such as displacement of the OR
group in carboxylic esters, or self-condensations, respectively, remain limited to a
minimum. Enolates derived from lactones, e.g., γ-butyrolactone, are extremely
unstable and must be generated and functionalized below -70 °C [7].

Alkali amides, MNH_2, in situ prepared in liquid ammonia, are efficient
deprotonating agents in reactions with ketones, carboxylic *t*-butyl esters, *N,N*-
dialkylcarboxamides, and the analogous sulfur compounds. Primary-alkyl
carboxylic esters RR'CHCOOR'' and the sulfur analogues RR'CHCOSR'' and
RR'CHCSOR'' (R'' is prim.-alkyl), however, undergo serious competitive
substitution of OR'' or SR'' by NH_2 [6].

Alkali amides seem unsuitable for the generation of 'kinetic' enolates from non-
symmetrical ketones, since in the presence of ammonia a ready transformation (via
proton donation and abstraction) into the 'thermodynamic' enolate can occur.

Literature

1. Weiler L (1970) J Am Chem Soc 92:6702
2. Groenewegen P, Kallenberg H, van der Gen A (1978) Tetrahedron Lett 491
3. Pi R, Friedl T, and von R Schleyer P, Klusener PAA, Brandsma L (1987) J Org Chem 52:4299
4. House HO, Czuba LJ, Gall M, Olmstead HD (1969) J Org Chem 34:2324
5. d'Angelo J (1976) Tetrahedron 32:2979
6. Brandsma L (unpublished)
7. Posner GH, Loomis GL (1972) J Chem Soc Chem Comm 892

3 Other Methods for the Generation of Enolates and Enethiolates

A useful method for preparing specifically substituted enolates consists in reducing an α,β-unsaturated ketone with lithium in liquid ammonia in the presence of a suitable proton donor, such as t-butylalcohol [1, 2], e.g.:

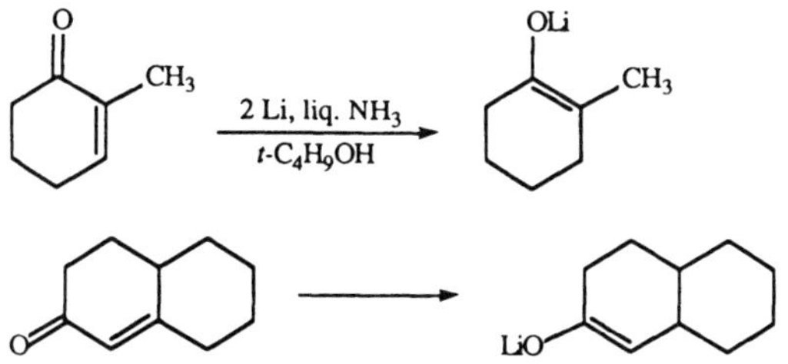

As mentioned in Chap. I, Section 13, interaction between α,β-unsaturated carbonyl compounds (C=C—C=O) and polar organometallic derivatives may give rise to a clean 1,4-addition resulting in an enolate R—C—C=C—OM. This method suffers from the lack of generality, however.

Trimethylsilyl enol ethers are converted into lithium enolates and tetramethylsilane by reaction with methyllithium [3]. This cleavage reaction not only provides a possibility of generating substituted enolates but also constitutes a method for preparing solutions of lithium enolates that are free from diisopropylamine, which compound may interfere in certain functionalization reactions. We found that the cleavage of enol ethers RCH=CHOSiMe₃ proceeds equally well with n-BuLi in an Et₂O-hexane mixture [6]. THF undergoes a cycloeliminative degradation upon reaction with BuLi [4] at temperatures above 20 °C, affording ethene and the enolate of acetaldehyde:

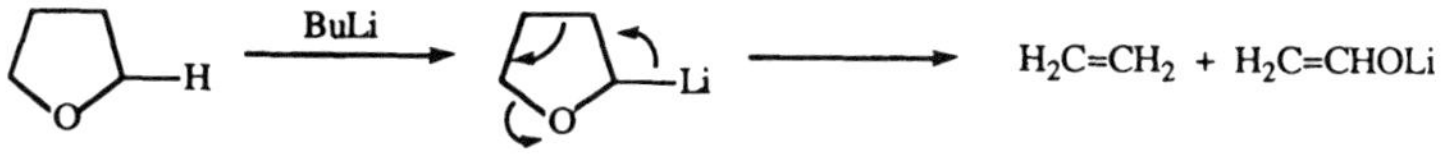

1-Alkenyl sulfides undergo reductive cleavage by alkali metals in liquid ammonia [5], e.g.,

$$H_2C{=}CHSC_2H_5 \xrightarrow[\text{liq. } NH_3]{2Li} H_2C{=}CHSLi + LiNH_2 \xrightarrow[\text{(1 equiv.)}]{NH_4Cl}$$

$$H_2C{=}CHSLi + LiCl + NH_3$$

Literature

1. Stork G, Darling SD (1964) J Am Chem Soc 86:1761
2. d'Angelo J (1976) Tetrahedron 32:2979
3. House HO, Czuba LL, Gall M, Olmstead HD (1969) J Org Chem 34:2324; Stork G, Ganem B (1973) J Am Chem Soc 95:6152
4. Jung ME, Blum RB (1977) Tetrahedron Lett 3791
5. Brandsma L (1988) Preparative acetylenic chemistry, Revised edition, Elsevier, p 271
6. Brandsma L, Verkruijsse HD (unpublished)

4 Dimetallation of Carbonyl and Thiocarbonyl Compounds

Symmetrical ketones like acetone and cyclohexanone have been converted into 'dianions' by successive treatment with potassium hydride and BuLi.TMEDA [1] e.g.:

$$CH_3C(O)CH_3 \xrightarrow{KH} {}^-CH_2C(O)CH_3 \xrightarrow[Et_2O]{BuLi \cdot TMEDA} {}^-CH_2C(O)CH_2{}^-$$

In contrast to the mono-anions from ketones, the dianions are powerful C-nucleophiles reacting vigorously with alkyl halides RX to give the enolates, e.g., $RCH_2C({=}O)CH_2{}^-$. We were unable to substantiate the good results reported.

Reaction of the (thermodynamic) enolate $PhCH{-}COCH_3$ (generated from phenylacetone and sodium hydride) with BuLi in a THF-hexane mixture gives a dark-coloured solution of the dianion $PhCH{-}COCH_2^-$ [2, 3]. Due to competitive attack by BuLi on the solvent THF, the second metallation is not complete, however [4].

Allylic ketones, e.g., $t{-}C_4H_9COCH_2CH{=}CHCH_3$, and allylic dithioesters, $H_2C{=}CHCH_2CH_2CSSR$, have been converted into the dimetallated compounds $(t{-}C_4H_9COCHCH{=}CHCH_2$ and $H_2C{=}CHCHCH{=}C(S^-)SR$, respectively) by successive treatment with KH and sec-BuLi·TMEDA in hexane [11]. With one equivalent of an electrophile (e.g., $Ph_2C{=}O$), the species react exclusively or predominantly on the remote carbon atom. After hydrolytic work-up, the products $E{-}CH_2CH{=}CHCH_2COt{-}C_4H_9$ and $E{-}CH_2CH{=}CH{-}CH_2CSSR$ are obtained.

Treatment of 1, 3-diketones, such as $CH_3COCH_2COCH_3$, with two equivalents of sodamide or potassium amide in liquid ammonia followed by reaction with alkyl halides gives the homologues in good yields [5, 6]; e.g.:

$$CH_3COCH_2COCH_3 \xrightarrow{^-NH_2} CH_3COCHCOCH_3 \xrightarrow{^-NH_2}$$

$$^-CH_2COCHCOCH_3 \xrightarrow{RX} RCH_2COCHCOCH_3 \xrightarrow{H^+, H_2O}$$

$$RCH_2COCH_2COCH_3$$

The alkylation proceeds exclusively on the most strongly basic carbon atom. In the cases of 1,3-diketones, the dianion formation is presumably complete. Dimetallation can also be achieved with two equivalents of LDA in THF. Alkali amides have been found to be unsuitable for the dimetallation of ketoesters, CH_3COCH_2COOR [5]. Possible reasons are incomplete second metallation and competitive attack by $^-NH_2$ on COOR. Although double metallation of these ketoesters with NaH and BuLi or with two equivalents of LDA and subsequent γ-alkylation has been reported to give high yields [6,7], our results were moderate.

Carboxylic acids can be successively metallated on O and on the α-carbon atom [8, 9]:

$$RCH_2COOH \xrightarrow{\text{NaH}} RCH_2COONa \xrightarrow{\text{LDA}} RCH(Li)COONa$$

$$RCH_2COOH \xrightarrow{\text{2LDA}} RCH(Li)COOLi$$

As expected, reaction of the dimetallated acids with one equivalent of electrophilic reagents occurs specifically on the strongly basic C-centre. With acetic acid we obtained poor results in alkylation reactions (compare Refs. [9] and [10]), presumably due to the slight solubility of the 'dianions'. Carboxylic acids with a longer carbon chain give the alkylation products in good yields [10].

Literature

1. Hubbard JS, Harris TM (1980) J Am Chem Soc 102:2110
2. Mao C, Hauser CR, Miles ML (1967) J Am Chem Soc 89:5303
3. Philip Bays J (1978) J Org Chem 43:38
4. Verkruijsse HD, Brandsma L (unpublished)
5. Harris TM, Harris CM (1969) Org Reactions 17:155, John Wiley
6. Kaiser EM, Petty JD, Knutson PLA (1977) Synthesis 509
7. Weiler L (1970) J Am Chem Soc 92:6702
8. Ivanov D, Vassilev G, Panayotov I (1975) Synthesis 83
9. Creger PL (1972) J Org Chem 37:1907; (1970) J Am Chem Soc 92:1397
10. Pfeffer PE, Silbert LS, Chirinko JM (1972) J Org Chem 37:451
11. Seebach D, Pohmakotr M (1981) Tetrahedron 37:4047

5 Experiments

All temperatures are internal, unless indicated otherwise.
All reactions are carried out in an atmosphere of inert gas.
For general instructions concerning handling alkali metal reagents, drying solvents, etc., see Vol. 1, Chap. 1.

5.1 Conversion of Ketones into Lithium Enolates and Subsequent Trimethylsilylation (General Procedure)

(Analogous reactions with other ketones)

Apparatus: p. 24, Fig. 1; 500 ml

Scale: 0.1 molar

Introduction

In their paper on the preparation of trimethylsilyl enol ethers, House et al. [1] illustrate the generation of lithium enolates with the reaction between 2-methylcyclohexanone and LDA using 1,2-dimethoxyethane as a solvent. The authors do not indicate a temperature range for the lithiation of the ketone, but we found the deprotonation of a variety of ketones to proceed extremely fast and with excellent results at temperatures in the region of $-70\,°C$. In view of the possibility of aldol condensation, dropwise addition of the ketone to the LDA solution seems advisable. Attempts to prepare enolates from aldehydes $RCH_2CH{=}O$ and LDA gave, after quenching with trimethylchlorosilane, only unidentified resinous products.

Comparision of our results with those reported by House and co-workers suggests that the solvent may influence the ratio of the two enolates ('kinetic' and 'thermodynamic') formed from non-symmetrical ketones, Whereas House et al. obtained exclusively the most stable enolate $PhCH{=}C(OLi)CH_3$ from LDA and $PhCH_2COCH_3$ in 1,2-dimethoxyethane, we found comparable amounts of $PhCH{=}C(OLi)CH_3$ $(E+Z)$ and $PhCH_2C(OLi){=}CH_2$ in our experiments carried out with LDA in 1:1 mixture of THF and hexane. Using the method of House, $CH_3COC_4H_9$ gives a 86:14 mixture of $H_2C{=}C(OLi)C_4H_9$ and $CH_3C(OLi){=}CHC_3H_7$. Metallation of $CH_3COC_9H_{19}$ with LDA in THF gave exclusively $H_2C{=}C(OLi)C_9H_{19}$ [2].

In their procedure for the preparation of trimethylsilyl enol ethers, House et al. use a very large excess of Me_3SiCl. Triethylamine was introduced together with Me_3SiCl, but the amount of this amine mentioned in their experimental procedures is not sufficient to neutralize all hydrogen chloride liberated during the aqueous work-up. Although the authors use an aqueous solution of $NaHCO_3$, their procedure involves the risk of hydrolytic cleavage in the case of sensitive enol ethers. We use a smaller excess of Me_3SiCl and neutralize this by addition of diethylamine $((C_2H_5)_2NH)$ prior to the aqueous work-up.

Procedure

A mixture of 0.11 mol (11.1 g) of diisopropylamine and 70 ml of THF is cooled to below $-50\,°C$. A solution of 0.11 mol of butyllithium in 73 ml of hexane is added over a few s by syringe. The resulting solution of LDA is cooled to $\sim -75\,°C$, after which a mixture of 0.1 mol (11.2 g) of 2-methylcyclohexanone and 20 ml of THF is added dropwise over 10 min. During this addition the reaction mixture is kept between -70 and $-60\,°C$. The cooling bath is then removed and the temperature of the faintly coloured solution allowed to rise to $-50\,°C$. Freshly distilled Me_3SiCl (0.13 mol, 14.1 g) is added in one portion, while allowing the temperature to rise to $\sim 15\,°C$. Diethylamine (3 g) is subsequently added to the white suspension and stirring at $10-20\,°C$ is continued for 5 min (note 1). The reaction mixture is then poured into 200 ml of ice water. After vigorous shaking and separation of the layers, the aqueous layer is extracted twice with small portions of pentane. The combined organic solutions are washed once with a concentrated aqueous solution of ammonium chloride and dried over $MgSO_4$. The solution is concentrated under reduced pressure and the remaining liquid carefully distilled to give the pure silyl enol ether b.p. $70\,°C/12$ mmHg, $n_D(20)$ 1.4453, in greater than 80% yield. The following compounds were prepared in a similar way (yields $> 75\%$):

$H_2C{=}C(OSiMe_3)C_9H_{19}$;   b.p.   124 °C/12   mmHg, $N_D(20)$ 1.4332,   from $H_3CCOC_9H_{19}$;

—OSiMe₃, b.p. 60°C/12 mmHg, n_D(20) 1.4460, from cyclohexanone;

$H_2C{=}C(OSiMe_3)Ph$, b.p. 93°C/15 mmHg, $n_D(20)$ 1.5018, from CH_3COPh;

—OSiMe₃, b.p. 60°C/12 mmHg, n_D(20) 1.4601, from

=O;

$H_2C{=}C(OSiMe_3)C{\equiv}CC_6H_{13}$, b.p. 115°C/14 mmHg, $n_D(20)$ 1.4542, from $CH_3COC{\equiv}CC_6H_{13}$ [3];

, b.p. ~110°C/1 mmHg, n_D(20) 1.4966, from

Notes

1. In the case of α,β-unsaturated ketones triethylamine is used.

Literature

1. House HO, Czuba LJ, Gall M, Olmstead HD (1969) J Org Chem 34:2324
2. Tip L, Andringa H, Brandsma L (unpublished)
3. Brandsma L (1988) Preparative acetylenic chemistry, Revised edition, Elsevier, p 103

5.2 Reaction of 1-Trimethylsilyloxy-1-Heptene with Butyllithium

$$C_5H_{11}CH=CHOSiMe_3 + BuLi \xrightarrow[0 \to 35\,°C]{Et_2O\text{-hexane}}$$

$$C_5H_{11}CH=CHOLi + BuSiMe_3$$

$$C_5H_{11}CH=CHOLi + (CH_3CO)_2O \xrightarrow[-80 \to -10\,°C]{}$$

$$C_5H_{11}CH=CHOCOCH_3 + CH_3COOLi$$

Apparatus: p. 24, Fig. 1; 11

Scale: 0.1 molar

Introduction

Direct α-metallation of aldehydes has been achieved by van der Gen and co-workers using commercially available potassium hydride [1]. We prepared [2] superactive potassium hydride by reaction of hydrogen with a 1:1:1 molar mixture of BuLi, t-BuOK, and TMEDA in hexane. Reaction at low temperatures with aliphatic aldehydes gave the potassium enolates RCH=CHOK with a selectivity of $\sim 90\%$ (10% of alcoholate RCH$_2$CH$_2$OK was formed by reduction). Both methods have their drawback: removal of the protecting paraffin oil from the commercially available KH and subsequent weighing involves the risk of fire hazards, while the presence of t-BuOLi and TMEDA in our suspension of KH may give rise to undesired reactions when the solution of potassium enolate is allowed to react with electrophilic reagents, such as sulfenyl halides [3] or carboxylic anhydrides.

LDA is totally unsuitable for the generation of enolates from aldehydes: quench reactions of the resulting mixtures with chlorotrimethylsilane gave only unidentified viscous products [4].

The reaction of the readily accessible trimethylsilyl enol ethers with methyllithium has been successfully applied to prepare ethereal solutions of lithium alkenolates RCH=CHOLi [3,5]. Good results can be obtained also with n-BuLi [4]. Both the trimethylsilyl enol ethers, which are prepared by heating a mixture of the aldehyde, chlorotrimethylsilane, triethylamine and DMF, and the enolates are mixtures of the E- and Z-isomers.

Procedure

A solution of 0.105 mol of butyllithium in 68 ml of hexane is added (by syringe) over a few min to a mixture of 0.1 mol (18.6 g) of 1-trimethylsilyloxy-1-heptene (E- and Z-isomer, see Exp. 12) and 80 ml of Et$_2$O, while keeping the temperature of the solution between 0 and 10 °C. After the addition, the cooling bath is removed so that the temperature can rise above 30 °C. After the temperature has dropped to 30 °C, the almost colourless solution is warmed for an additional 30 min at ~ 35 °C. Subsequent aqueous hydrolysis gives a colourless viscous liquid, presumably the aldol-condensation product, the starting compound being completely absent. Reaction at $0 \rightarrow 25$ °C of the solution of the enolate with a 10% molar excess of Me_3SiCl, followed by a work-up as described in Exp. 1, gives the original silyl enol ether in ~ 85% yield, only a small viscous residue remaining after distillation.

To prepare the enol acetate pure acetic anhydride (0.11 mol, 11.2 g) is added in one portion to the ethereal solution of lithium heptenolate cooled at − 80 °C. A gelatinous suspension of lithium acetate is formed immediately. After the addition, the cooling bath is removed and the temperature allowed to rise to − 20 °C. Ice water (100 ml) is then added with vigorous stirring. The upper layer and one ethereal extract are combined and dried over $MgSO_4$. After removal of the solvent in vacuo the remaining liquid is distilled through a 20-cm Vigreux column to give the enol ester, b.p. 75 °C/12 mmHg, $n_D(20)$ 1.4318, in ~ 75% yield. The E/Z-ratio is the same as that of the starting compound (~ 3:2).

Literature

1. Groenewegen P, Kallenberg H, van der Gen A (1978) Tetrahedron Lett 491
2. Pi R, Friedl T, von R Schleyer P, Klusener PAA, Brandsma L (1987) J Org Chem 52:4299
3. Seebach D, Teschner M (1973) Tetrahedron Lett 5113
4. Brandsma L (unpublished)
5. Stork G, Ganim B (1973) J Am Chem Soc 95:6152

5.3 Metallation of Carboxylic Esters with LDA

$$RCH_2COOR' + LDA \xrightarrow[-70 \rightarrow -40\,°C]{THF\text{-}hexane} RCH(Li)COOR' + HDA$$

Reaction with allyl bromide, aldehyde, ketones, Me_3SiCl, I_2.

Apparatus: p. 24, Fig. 1; 11

Scale: 0.1 molar

Introduction

α-Lithiation of carboxylic esters has been achieved by Rathke using lithium isopropyl cyclohexylamide (LICA), lithium bis(trimethylsilyl)amide, or LDA in THF [1-3]. With these bulky reagents, attack on the C=O function with concomitant replacement of the OR group by NR_2 is insignificant. Another potential side-reaction is the self-condensation of the ester by the reaction of the initial enolate with the substrate. This process can be suppressed by using a kinetically active base. In this respect LDA and LICA are preferred over LiN(SiMe$_3$)$_2$, which is less strongly basic (pK HN(SiMe$_3$)$_2$ ~ 31) than HN(i-C$_3$H$_7$)$_2$ and HN(i-C$_3$H$_7$) (c-C$_6$H$_{11}$) (pK ~ 36). For similar reasons, THF is a better choice than the less polar Et$_2$O or hexane. Acetic esters are kinetically more acidic than their homologues and therefore the side-reactions mentioned are expected to compete less seriously. t-Butyl esters should be better substrates for metallations than primary-alkyl esters because O—t-C$_4$H$_9$ is a poorer leaving group.

Compared with several synthetically important anions, ester enolates are rather poor nucleophiles in conversions with alkyl halides and epoxides [4], reactions that have a relatively high activation energy barrier. Yields of the alkylation products are often rather low (especially in reactions with secondary alkyl bromides and iodides) due to the occurrence of condensation of the alkylation product with unreacted enolate. Improved results in alkylations may be obtained when using the very polar DMSO or HMPT as co-solvent [1]; under these conditions only C-alkylation products are formed.

Reaction of lithium enolates from primary alkyl esters with chloro-trimethylsilane occurs under mild conditions and affords the water-sensitive ketene-acetals RCH=C(OR′)(OSiMe$_3$). t-Butyl esters yield comparable amounts of the O and C-silylated derivatives [5].

Aldol condensations with aldehydes and ketones proceed smoothly at temperatures below − 60 °C and usually give high yields [3].

Carboxylic esters with a halogen atom next to the ester function are highly reactive alkylating agents. If iodine is added to a solution of the enolate, the initial α-iodoester reacts immediately with unconverted enolates to give the 'Würtz'-coupling product. High yields of α-iodoesters can be obtained, however, if inversed-order addition is applied [6].

Procedure

a) A solution of 0.105 mol of BuLi in 68 ml of hexane is added over a few s (by syringe) to a mixture of 0.11 mol (11.1 g) of diisopropylamine and 70 ml of THF cooled to − 70 °C. Subsequently, the ester (0.1 mol) is added dropwise over 15 min while the temperature is kept between − 65 and − 75 °C. After the addition, the cooling bath is removed and the temperature allowed to rise to − 40 °C.

b) The solution is cooled again to − 70 °C and 0.15 mol (18.0 g) of allyl bromide is added in one portion. The cooling bath is removed and the temperature allowed to rise to 20 °C. Stirring is continued for 1 h, then 100 ml of water is added with vigorous stirring. The aqueous layer is extracted twice with Et$_2$O. After drying, the combined organic solutions over MgSO$_4$ the solvent is removed under reduced pressure. If the expected b.p. of the product is lower than 70 °C/15 mmHg, the greater part of the solvent should be distilled off at normal pressure. The product is isolated by distillation in vacuo. H$_2$C=CHCH(C$_2$H$_5$)COOCH$_3$, b.p. 48 °C/12 mmHg, n$_D$(20) 1.4237, was obtained in ~ 70% yield. For reactions with saturated alkyl halides see Refs. [1,7] and Exp. 6.

c) The aldehyde or ketone (0.1 mol) is added over a few s to the solution of the lithium enolate cooled to $-70\,°C$. After 5 min, 100 ml of water is added with vigorous stirring and the product is isolated in the usual way. The following compounds were prepared (yields > 70%):

$PhCH(OH)CH_2COOC_2H_5$, b.p. $\sim 110\,°C/0.5$ mmHg, $n_D(20)$ 1.5130;
$PhCH(OH)CH(CH_3)COO—t-C_4H_9$, b.p. $\sim 120\,°C/0.5$ mmHg, $n_D(20)$ 1.4936;

$CH_2COOC_2H_5$, b.p. $115\,°C/14$ mmHg, $n_D(20)$ 1.4537.

d) Freshly distilled chlorotrimethylsilane (0.12 mol, 13.0 g) is added in one portion at $-70\,°C$ and the cooling bath is removed. The solution (at a later stage, white suspension) is warmed up to $+30\,°C$. Since the product is water-sensitive, a 'dry' work-up is carried out. The suspension is transferred into a 1-l round-bottom and the volatile components are removed under reduced pressure using a rotary evaporator. If the expected b.p. of the ketene acetal is lower than $60\,°C/15$ mmHg, some of the product may be swept along with the solvent distilled off. To minimize losses, the bath temperature should be kept below $35\,°C$ in these cases. The remaining liquid (ketene acetal and some hexane and THF) is distilled off from the salt under the lowest possible pressure: the 1-l flask is connected to a short (5–10 cm) Vigreux column, condenser, and *single* receiver immersed in a bath at $-78\,°C$. When the pressure has become minimal (< 1 mmHg), the 1-l flask is placed in a heating bath and the bath temperature gradually raised until only dry salt remains. Careful redistillation of the contents of the receiver gives the ketene acetal. $C_2H_5CH=C(OSiMe_3)OCH_3$, b.p. $55\,°C/15$ mmHg, $n_D(20)$ 1.4178, was obtained in 70–75% yield.

Starting from $C_2H_5COO—t$-Bu comparable amounts of $CH_3CH(SiMe_3)$ $COO—t$-Bu and $CH_3CH=C(OSiMe_3)O—t$-Bu were obtained.

e) Iodine (0.13 mol, 33.0 g) is dissolved in 110 ml of THF. The stirred solution is cooled to $\sim -90\,°C$ (occasional cooling in a bath with liquid N_2), after which the solution of the enolate (temperature $\sim -20\,°C$) is added over ~ 10 min (by syringe) with vigorous stirring. During this addition the temperature is allowed to rise gradually to $\sim -50\,°C$. Ten min after completion of the addition a solution of 15 g of $Na_2S_2O_3$ in 150 ml of water is added with vigorous stirring. The layers are separated and the aqueous layer is extracted twice with Et_2O. The combined organic solutions are dried over $MgSO_4$ and subsequently concentrated in vacuo (if the expected b.p. of the iodo ester is below $65\,°C/15$ mmHg, the bath temperature should be kept below $35\,°C$). Distillation of the remaining liquid gives the α-iodo ester.

$C_2H_5CH(I)COOCH_3$, b.p. $70\,°C/14$ mmHg, $n_D(20)$ 1.5016, was obtained in 74% yield.

Literature

1. Rathke MW, Lindert A (1971) J Am Chem Soc 93:2318
2. Rathke MW (1970) J Am Chem Soc 92:3222

3. Rathke MW, Sullivan DF (1973) J Am Chem Soc 95:3050
4. Sturm TJ, Marolewski AE, Rezenka DS, Taylor SK (1989) J Org Chem 54:2039
5. Brandsma L, Verkruijsse HD (unpublished)
6. Rathke MW, Lindert A (1971) Tetrahedron Lett 3995
7. Cregge RJ, Herrmann JL, Lee CS, Richman JE, Schlessinger RH (1973) Tetrahedron Lett 2425

5.4 Metallation of Carboxylic Esters and Amides in Liquid Ammonia

$$CH_3COO—t\text{-}C_4H_9 + LiNH_2 \xrightarrow[-70° \to -40\,°C]{\text{liq. }NH_3}$$

$$LiCH_2COO—t\text{-}C_4H_9\downarrow + NH_3$$

$$CH_3CONMe_2 + LiNH_2 \xrightarrow[-33\,°C]{\text{liq. }NH_3} LiCH_2CONMe_2 + NH_3$$

Alkylation with alkyl and benzyl halides.

Apparatus: p. 24, Fig. 1; 11

Scale: 0.3 molar

Introduction

Carboxylic esters undergo a facile α-deprotonation upon treatment with alkali amides in liquid ammonia. The obtained solutions react smoothly with alkyl halides to give the expected homologues [1]. Primary-alkyl esters, however, give strongly reduced yields of the alkylation product, due to competitive attack by $^-NH_2$ on the ester group with formation of carboxamides. t-Butoxy groups undergo this substitution to a lesser extent. This side reduction can be reduced further by maintaining a low temperature during the metallation reaction. Since $N(alkyl)_2$ is a poor leaving group, alkylation of N,N-dialkylcarboxamides in liquid ammonia can be accomplished with excellent results.

The alkylation in the polar liquid ammonia proceeds much faster than in THF (compare Refs. [2, 3]).

Procedure

Anhydrous liquid ammonia (0.5 l) is placed in the flask. A slow stream of nitrogen is passed through the flask and traces of water are 'neutralized' by introducing very small pieces (< 0.1 g each) of lithium until the blue colour persists. Iron(III)nitrate (~ 200 mg) is then added, followed by 0.35 mol (2.4 g) of lithium cut into pieces of ~ 0.5 g (for the preparation of $LiNH_2$ consult Vol. 1, Chap. I). When, after ~ 0.5 h, all lithium has been converted into lithium amide, the carboxylic amide (0.3 mol) is added over 10 min with vigorous stirring. After an additional 10 min, the alkylating agent (0.4 mol) is added over 5 min, while keeping the temperature between -50 and $-60\,°C$. In the case of carboxylic esters, the suspension of lithium amide is cooled down to $-70\,°C$ (bath with dry ice and acetone or with liquid nitrogen) while introducing nitrogen at a rate of $\sim 500\,ml/min$. The ester is then added over a few min. The temperature is allowed to rise to between -55 and $-50\,°C$. After an additional 15 min (at $\sim -50\,°C$), the alkylating agent (0.4 mol) is added over 10 min, while keeping the temperature between -40 and $-50\,°C$.

After the addition of the alkyl halides, the cooling bath is removed. Stirring is continued for 2 h, then the dropping funnel, thermometer, and outlet are removed and the flask is placed in a water bath at 40 °C. When the flow of escaping ammonia vapour has become very faint, 150 ml of water is added and the mixture is extracted with Et_2O: in the cases of carboxamides with a relatively low molecular weight, several extractions have to be carried out. The extracts are dried over $MgSO_4$, concentrated in vacuo and the products are isolated by distillation.

The following compounds were prepared:

$C_4H_9CH_2COO-t-C_4H_9$, b.p. 65 °C/12 mmHg, $n_D(20)$ 1.4106, yield 74%, from $CH_3COO-t-C_4H_9$, $LiNH_2$, and C_4H_9Br;

$PhCH_2CH_2COO-t-C_4H_9$, b.p. 128 °C/12 mmHg, $n_D(20)$ 1.4842 in 75% yield, from $CH_3COO-t-C_4H_9$, $LiNH_2$, and $PhCH_2Cl$;

$Cl(CH_2)_3CH_2COO-t-C_4H_9$, b.p. 100 °C/12 mmHg, $n_D(22)$ 1.4348, yield 70%, from $CH_3COO-t-C_4H_9$, $LiNH_2$, and $Cl(CH_2)_3Br$;

$(CH_3)_2CHCH_2CONMe_2$, b.p. 75 °C/12 mmHg, in 72% yield, from CH_3CONMe_2, $LiNH_2$, and $(CH_3)_2CHBr$;

$C_8H_{17}CH_2CONMe_2$, b.p. ~ 110 °C/0.4 mmHg, $n_D(19)$ 1.4543, in 82% yield from CH_3CONMe_2, $LiNH_2$, and $C_8H_{17}Br$;

$Cl(CH_2)_3CH_2CONMe_2$, b.p. 140 °C/15 mmHg, $n_D(23)$ 1.4747, in 75% yield from CH_3CONMe_2, $LiNH_2$, and $Cl(CH_2)_3Br$.

Literature

1. Hauser CR, Chambers WJ (1956) J Org Chem 21:1524
2. Rathke MW, Lindert A (1971) J Am Chem Soc 93:2318
3. Cregge RJ, Herrmann JL, Lee CS, Richman JE, Schlessinger RH (1973) Tetrahedron Lett 2425

5.5 Metallation of N-Methylpyrrolidinone and γ-Butyrolactone with LDA

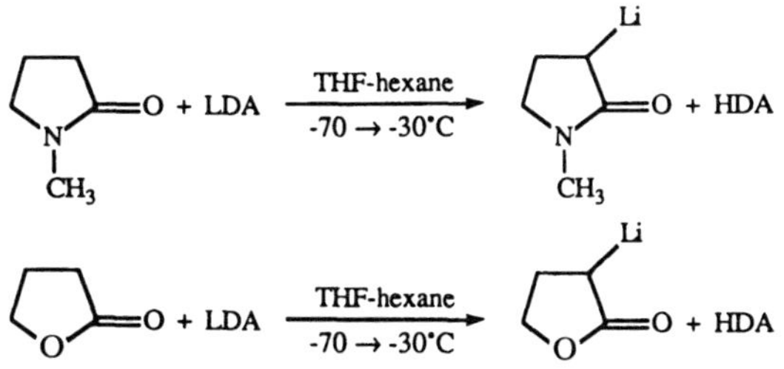

Alkylation with allyl bromide.

Apparatus: p 24, Fig. 1; 500 ml

Scale: 0.1 molar

Introduction

The conditions for the lithiation of N-methylpyrrolidinone described below, are representative for N,N-dialkylcarboxamides and illustrate the ready metallation of this class of compounds in THF-hexane mixtures. Solutions of lithiated carboxamides have a reasonable stability even at somewhat elevated temperatures.

Two very concise reports [1,2] on the lithiation of γ-butyrolactone and subsequent functionalization with reactive alkylating agents give contradictionary information about the stability of the lithiated lactone. Whereas Posner concludes from his deuteration experiments that the lithiated lactone undergoes appreciable decomposition at temperatures above − 70 °C, Schlessinger reports excellent yields of products obtained by reaction of the intermediate with a number of alkylating agents at − 30 °C. Our findings [3] are in agreement with those reported by Posner. A successful alkylation with the reactive halide allyl bromide could be achieved only, when HMPT was used as co-solvent and the temperature was kept very low. In the absence of HMPT, alkylation did not occur at all. The lithiated lactame reacted very smoothly at very low temperatures with allyl bromide in a mixture of THF and hexane.

Procedure

a) N-Methylpyrrolidinone (0.1 mol, 9.9 g, dissolved in 20 ml of THF) is added in one portion to a solution of 0.105 mol of LDA in 68 ml of hexane and 70 ml of THF (see Exp. 3a), cooled at − 70 °C. The cooling bath is removed and the temperature allowed to rise to − 30 °C. The clear, light-yellow solution is cooled again to − 90 °C and 0.15 mol (18.2 g) of allyl bromide is added in one portion, after which the cooling bath is removed. The temperature rises within a few s to above − 40 °C. The solution is warmed to 20 °C and subsequently transferred into a 1-l round-bottom. After the greater part of the solvent has been removed under reduced pressure, 60 ml of water is added. The mixture is extracted five times with Et_2O. After drying the (unwashed) extracts over $MgSO_4$, the solvent is removed in vacuo and the remaining liquid distilled through a 20-cm Vigreux column. The alkylation product, b.p. ~ 70 °C/0.4 mmHg, $n_D(20)$ 1.4810, is obtained in 84% yield.

N,N-Dimethylacetamide can be lithiated under similar conditions. Reaction of the enolate with reactive halides such as $PhCH_2Br$ and $H_2C=CHCH_2Br$ proceeds very smoothly at temperatures below 0 °C. Saturated alkyl bromides react at ambient or slightly elevated temperature. The products are conveniently isolated with high yields as described above. Condensation products of lithiated carboxamides and aldehydes or ketones are hydrolyzed and isolated by the special procedures described in Chap. VIII, Exp. 4 and Chap. VI, Exp. 2d.

b) γ-Butyrolactone (8.6 g, 0.10 mol), diluted with 20 ml of THF, is added dropwise over 15 min to a solution of 0.11 mol of LDA in 74 ml of hexane and 70 ml of THF, while keeping the temperature around $-90\,°C$. The temperature is then allowed to rise over 10 min to $-80\,°C$, after which 0.15 mol (18.2) of allyl bromide is added over 10 min to the colourless solution, whilst maintaining the temperature between -80 and $-85\,°C$. Dry HMPT (25 ml, for drying HMPT see Vol. 1, Chap. I and p. 225) is then added over 10 min. The solution is stirred for an additional half hour at $-80\,°C$, then the cooling bath is removed and the temperature allowed to rise to $0\,°C$. The reaction mixture is transferred into a 1-l round-bottom, after which the THF and hexane are removed under reduced pressure, using a rotary evaporator. Water (100 ml) is added to the remaining viscous liquid. The mixture is extracted eight times with small portions (2×70 ml and 5×30 ml) of Et_2O. Each etheral extract is washed twice with 30 ml portions of water, the washings being added to the first aqueous layer. The combined ethereal solutions are dried over $MgSO_4$, after which the allylation product, b.p. $102\,°C/12$ mmHg, $n_D(20)$ 1.4618, is isolated in 68% yield.

Literature

1. Posner GH, Loomis GL (1972) J Chem Soc Chem Comm 892
2. Herrmann JL, Schlessinger RH (1973) J Chem Soc Chem Comm 711
3. Verkruijsse HD, Brandsma L (unpublished)

5.6 Lithiation of Methyl Crotonate

$$CH_3CH{=}CHCOOCH_3 + LDA \xrightarrow[-90\,°C]{THF\text{-}hexane\text{-}HMPT}$$

$$H_2C{=}CHCH(Li)COOCH_3 + HDA$$

$$H_2C{=}CHCH(Li)COOCH_3 + C_4H_9I \longrightarrow$$

$$H_2C{=}CHCH(C_4H_9)COOCH_3 + LiI$$

Apparatus: p. 24, Fig. 1; 11

Scale: 0.1 molar

Introduction

The HMPT used as a co-solvent in the procedure described below [1] has a double function. Reaction of methyl crotonate with LDA in a THF-hexane mixture gives rise to predominant conjugate addition of the base. HMPT causes such an enormous increase of the kinetic basicity of LDA, that abstraction of a proton becomes the only reaction. Since ester enolates are relatively weak nucleophiles in reactions with alkyl halides, elevated temperatures are needed to achieve alkylation in the organic solvents. Under these conditions, side-reactions such as condensation are likely to occur. HMPT is known to cause a dramatic increase of the reactivity of polar organometallic compounds especially in alkylations. Thus, these reactions can take place at an acceptable rate at much lower temperatures.

Procedure

A solution of 0.1 mol of LDA in 66 ml of hexane and 70 ml of THF (see Exp. 1) is cooled to below $-60\,°C$ after which 20 ml HMPT (for purifying the commercial solvent see Vol. 1, Chap. I and p. 225) is added over a few min. A mixture of 0.12 mol (12.0 g) of methyl crotonate and 20 ml of THF is added dropwise over 10 min, while maintaining the temperature between -80 and $-90\,°C$ (occasional cooling in a bath with liquid N_2). After an additional 5 min, 0.1 mol (18.4 g) of butyl iodide is added over 10 min at -80 to $-90\,°C$. The cooling bath is then removed and the temperature allowed to rise to $-10\,°C$. The mixture is stirred for an additional half hour at 20 °C, then a mixture of 12 g of 37% aqueous hydrochloric acid and 100 ml of water is added with vigorous stirring. After separation of the layers, one extracion with 50 ml of Et_2O is carried out. The organic solution is washed three times with a saturated aqueous solution of ammonium chloride, then dried over $MgSO_4$, and concentrated under reduced pressure. The alkylation product, b.p. 67 °C/12 mmHg, $n_D(23)$ 1.4267, is obtained in 70% yield. The residue consists mainly of the disubstitution product, $H_2C=CHC(C_4H_9)_2COOCH_3$.

Literature

1. Herrmann JL, Kieczykowski GR, Schlessinger RH (1973) Tetrahedron Lett 2453

5.7 Lithiation of Thiolesters, Thionesters, and Dithioesters with LDA (General Procedure)

$$RCH_2C(=O)SR' + LDA \xrightarrow[-75\,°C]{\text{THF-hexane}} RCH(Li)C(=O)SR' + HDA$$

$$RCH_2C(=S)SR' + LDA \xrightarrow{\text{idem}} RCH(Li)C(=S)SR' + HDA$$

$$RCH_2C(=S)OR' + LDA \xrightarrow{\text{idem}} RCH(Li)C(=S)OR' + HDA$$

$$(R = H \text{ or alkyl}; R' = alkyl)$$

Functionalization reactions with allyl bromide, pivaldehyde, and chlorotrimethylsilane.

Apparatus: p. 24, Fig. 1; 500 ml

Scale: 0.1 molar

Procedure

a) The carbonyl or thiocarbonyl compound (0.1 mol, diluted with 20 ml of THF) is added dropwise over 10 min to a solution of 0.10 mol of LDA (Exp. 1) in 66 ml of hexane and 70 ml of THF kept between -70 and $-80\,°C$. After an additional 10 min functionalization reactions are carried out.

b) Allyl bromide (0.12 mol, 14.5 g) is added in one portion to the (colourless) solution of $CH_3CH(Li)C(=S)OCH_3$, the cooling bath is removed and the solution warmed to $+30\,°C$. Under these conditions the initial S-alkylation product undergoes a 3,3-sigmatropic rearrangement [1] to give $H_2C=CHCH_2CH(CH_3)C(=S)OCH_3$. The reaction mixture is hydrolyzed with 100 ml of ice water and, after one extraction with Et_2O, the organic solution is dried over $MgSO_4$. The greater part of the solvent is distilled off at normal pressure through a 30-cm Vigreux column. Since thionesters are oxygen-sensitive, the distillation is carried out under nitrogen. After cooling to room temperature, the remaining liquid is carefully distilled to give the alkylation product, b.p. $50\,°C/13\,mmHg$, $n_D(20)$ 1.4794, in 77% yield.

c) Pivaldehyde (0.1 mol, 8.6 g) is added in one portion at $-70\,°C$ to the solution of $LiCH_2C(=O)SC_2H_5$ or $LiCH_2C(=S)SC_2H_5$, whereupon the cooling bath is removed. The solution is warmed to $0\,°C$, after which 100 ml of water is added with vigorous stirring. After separation of the layers and two extractions with Et_2O, the organic solution is dried over $MgSO_4$ and concentrated under reduced pressure. Distillation of the remaining liquid through a 20-cm Vigreux column gives the carbinol $t\text{-}C_4H_9CH(OH)CH_2C(=O)SC_2H_5$, b.p. $\sim 60\,°C/0.2\,mmHg$, $n_D(20)$ 1.4780 in 74% yield and the α,β-unsaturated dithioester $t\text{-}C_4H_9CH=CHC(=S(SC_2H_5$ (E-isomer, formed by elimination of water from the carbinol during the distillation), b.p. $\sim 70\,°C/0.15\,mmHg$, $n_D(21)$ 1.5612, in $\sim 55\%$ yield.

d) Freshly distilled chlorotrimethylsilane (0.11 mol, 11.9 g) is added in one portion at $-70\,°C$ to the solution of the lithiated dithioester. The reaction mixture is warmed to room temperature, after which it is transferred into a 1-1 round-bottom. The solvent is removed under reduced pressure, after which the product is isolated as described in exp. 3d. $C_2H_5CH=C(SSiMe_3)(SCH_3)$, b.p. $52\,°C/0.4\,mmHg$, $n_D(20)$ 1.5175, is obtained in $\sim 70\%$ yield. For other examples see Ref. [2].

Literature

1. Schuijl PJW, Brandsma L (1968) Recl Trav Chim Pays-Bas 87:929
2. Sukhai SR, Brandsma L (1979) Synthesis 455

5.8 Metallation of Dithioesters with Alkali Amide in Liquid Ammonia

$$RCH_2CSSR' + MNH_2 \xrightarrow[-33\,°C]{liq.\ NH_3} RCH{=}C(SR')SM + NH_3$$

S-alkylation with alkyl halides.

Apparatus: 1-l three-necked, round-bottomed flask equipped with a dropping funnel, a mechanical stirrer and an outlet (rubber stopper with a hole of ~ 7 mm diameter).

Scale: 0.2 molar

Introduction

As mentioned in Exp. 4, formation of enolates from carboxylic esters and alkali amides in liquid ammonia may proceed unsatisfactorily due to competitive attack on the C=O function by $^-NH_2$. This reaction is also observed in the cases of thiol esters $RCH_2C({=}O)SR'$ and, to a lesser extent, with thiono esters, $RCH_2C({=}S)OR'$ [1]. With the more acidic dithioesters, $RCH_2C({=}S)SR'$, this reaction is in most cases insignificant. Reaction of the ammoniacal solutions, obtained from dithioesters and alkali amides, with alkyl bromides or iodides generally affords exclusively the S-alkylation products. If the alkylation is performed with an allylic halide, the initial keten-S,S-acetal undergoes a 3,3-sigmatropic rearrangement during the work-up or during distillation to give the C-allylation product.

Procedure

Anhydrous liquid ammonia (350 ml) is placed in the flask. Traces of water are first 'neutralized' by adding — with intervals — very small pieces of sodium, until the blue colour persists. Subsequently, 0.22 mol of sodamide is prepared by successively introducing ~ 200 mg of iron(III)nitrate and 5.1 g of sodium cut in pieces of ~ 0.5 g (see Vol. 1, Chap. I). The dithioester (0.2 mol, diluted with some Et_2O or THF) is then added over 10 min with vigorous stirring. The addition is most conveniently carried out by syringe, taking care that the dithioesters does not run down the glass wall (this might give rise to some self-condensation). The conversion into the enethiolate is usually extremely fast, so that the alkyl bromide can be added immediately after the introduction of the dithioester. Volatile alkyl halides, such as ethyl bromide, are added in about 20% molar excess. Methyl iodide should be introduced by syringe directly into the solution to minimize attack by ammonia vapour.

After an additional 15 min, the ammonia is removed by placing the flask in a water bath at ~ 30 °C. For the work-up see Ref. [2]. Yields of the keten-S,S-acetals are mostly excellent.

Literature

1. Schuijl PJW, Brandsma L (1968) Recl Trav Chim Pays-Bas 87:929
2. Schuijl PJW, Brandsma L, Arens JF (1966) ibid. 85:1263

5.9 Dimetallation of Carboxylic Acids

$$RCH_2COOH + NaH \xrightarrow[0 \to 55\,°C]{THF} RCH_2COONa\downarrow + H_2\uparrow$$

$$RCH_2COONa + LDA \xrightarrow[-20 \to +30\,°C]{THF\text{-}hexane} RCH(Li)COONa\downarrow + HDA$$

$$RCH(Li)COONa + H_2C{=}CHCH_2Br \xrightarrow[0 \to 45\,°C]{THF\text{-}hexane}$$

$$H_2C{=}CHCH_2CH(R)COONa + LiBr \xrightarrow{H^+, H_2O}$$

$$H_2C{=}CHCH_2CH(R)COOH$$

Apparatus: p. 24, Fig. 1; 11

Scale: 0.1 molar

Introduction

The sodium salts of carboxylic acids, which can be readily prepared from the acids and sodium hydride in THF, can be deprotonated at the α-position by LDA. Alkylating reagents react specifically at the most strongly basic carbon atom. Since the carboxylic group is protected by its conversion into the sodium salt, the chance of a subsequent reaction of the α-alkylated salt with the dimetallated acid is small. For this reason, this method for homologation of carboxylic acid derivatives should be preferred to alkylation via carboxylic ester enolates. However, the slight solubility of the dimetallated acids in THF-hexane mixtures reduces their reactivity towards alkylating agents considerably. Especially dimetallated acetic acid was found to react very sluggishly with alkyl bromides and the alkyl derivatives were obtained in low yields. Addition of HMPT causes solubilization and led to improved results [1–4].

In our procedure α-lithiated sodium caproate is alkylated with the reactive allyl bromide, in which case the assistance of HMPT is not required.

Procedure

After having replaced the air in the flask completely by nitrogen, 0.12 mol of sodium hydride protected by mineral oil (content of NaH ~ 60 or 80%) is transferred into the flask applying a Schlenk technique. THF (100 ml) and diisopropylamine (0.1 mol, 10.1 g) are successively added. A mixture of 20 ml of THF and 0.1 mol (11.6 g) of dry caproic acid is added over 15 min while cooling the flask in a bath with ice water. After the evolution of hydrogen has ceased, the mixture is warmed to $\sim 50\,°C$ for 15 min. The thick suspension is then cooled to $-20\,°C$ (vigorous stirring is necessary) and a solution of 0.1 mol of BuLi in 66 ml of hexane is added (by syringe) over a few min, keeping the temperature below $+10\,°C$. The α-lithiation is brought to completion by stirring the suspension for an additional half hour at $35\,°C$. After cooling to $0\,°C$, 0.12 mol (14.5 g) of allyl bromide is added over 15 min, while keeping the temperature between 0 and $15\,°C$. The cooling bath is then removed and the suspension is heated for an additional half hour at $45\,°C$. Water (50 ml) is then added with vigorous stirring and cooling in a bath with ice water. Dilute hydrochloric acid (4 N) is then added dropwise with stirring until the aqueous layer has reached pH 2. The layers are separated and three extractions with Et_2O are carried out. The combined organic solutions are washed with a saturated aqueous solution of sodium chloride, dried over $MgSO_4$, and concentrated under reduced pressure. Distillation of the remaining liquid through a short column gives the desired carboxylic acid, b.p. $\sim 85\,°C/0.7\,mmHg$, $n_D(20)$ 1.4392, in $\sim 70\%$ yield.

Literature

1. Creger PL (1967) J Am Chem Soc 89:2500
2. Creger PL (1970) J Am Chem Soc 92:1397
3. Creger PL (1970) J Org Chem 37:1907
4. Pfeffer PE, Silbert LS, Chirinko JM (1972) J Org Chem 37:451

5.10 Dimetallation of Pentane-2,4-dione with Sodamide in Liquid Ammonia

$$CH_3COCH_2COCH_3 + 2NaNH_2 \xrightarrow[-33\,°C]{liq.\,NH_3}$$

$$NaCH_2COCH(Na)COCH_3$$

$$NaCH_2COCH(Na)COCH_3 + C_3H_7Br \longrightarrow$$

$$C_3H_7CH_2COCH(Na)COCH_3 + NaBr \xrightarrow{H^+,H_2O}$$

$$C_3H_7CH_2COCH_2COCH_3$$

Apparatus: 2-l round-bottomed, three-necked flask, equipped with a dropping funnel, a mechanical stirrer, and a gas-outlet (rubber stopper with a hole of $\sim 7\,mm$ diameter).

Scale: 0.3 molar

Procedure

(Compare Hampton KC, Harris TM, Hauser CR (1967) Org Syntheses 47:92)
Anhydrous liquid ammonia ($\sim$ 700 ml) is placed in the flask. Small amounts of water
are first neutralized by addition of small pieces ($\sim$ 0.1 g) of sodium until the blue
colour of dissolved sodium persists. The second and—if necessary—following
pieces are introduced not until the colour caused by the preceding one has
disappeared. Iron(III)nitrate ($\sim$ 300 mg) is added, after 20 s followed by $\sim$ 2 g of
sodium cut in pieces of $\sim$ 0.5 g. The remaining sodium pieces of the total amount of
0.6 mol (13.8 g) are added after the first 2 g are converted into sodamide (greyish
suspension). When the conversion of all sodium is complete, 100 ml of Et_2O is
added, after which 0.3 mol (30.0 g) of pentanedione is introduced over $\sim$ 10 min with
efficient stirring. The addition of the diketone is best carried out by means of a
syringe, the needle of which is held only a few cm above the suspension of sodamide.
For this operation the outlet is temporary removed.

Propyl bromide (0.45 mol, 55.3 g) is added dropwise over 30 min to the greyish
suspension. After an additional 1 h, the dropping funnel and outlet are removed and
the flask is placed in a water bath at 40 °C. When the flow of escaping ammonia
vapour has becomes faint, 250 ml of Et_2O is added and heating in a bath at $\sim$ 50 °C
is continued for an additional 15 min. A mixture of 40 g of 37% aqueous
hydrochloric acid and 250 ml of ice water is added while cooling the flask in a bath
with ice water. Then an additional amount of dilute HCl required to bring the
aqueous phase to $\sim$ pH 5 is gradually added. Four extractions with Et_2O are carried
out. The unwashed organic solutions are dried over $MgSO_4$, concentrated in vacuo,
and the remaining liquid is distilled to give octane-2,4-dione, b.p. 78 °C/12 mmHg
$n_D(20)$ 1.4596, in $\sim$ 90% yield.

Literature

See the reviews indicated by Refs. [9 and 11] in Sect. 1.

5.11 Dimetallation of Pentane-2,4-dione with LDA

$$CH_3COCH_2COCH_3 + 2LDA \xrightarrow[-20 \to +10\,°C]{\text{THF-hexane}}$$

$$LiCH_2COCH(Li)COCH_3 + 2HDA$$

$$LiCH_2COCH(Li)COCH_3 + H_2C{=}CHCH_2Br \xrightarrow[0 \to 30\,°C]{}$$

$$H_2C{=}CHCH_2CH_2COCH(Li)COCH_3 \xrightarrow{H^+, H_2O}$$

$$H_2C{=}CHCH_2CH_2COCH_2COCH_3$$

Apparatus: p. 24, Fig. 1; 500 ml

Scale: 0.05 mol

Introduction

Alkali enolates derived from ketones are rather weak nucleophiles in reactions with alkyl halides. In THF-hexane mixtures, the reactions with saturated alkyl halides proceed sluggishly and the yields are generally unsatisfactory. We showed that even in the polar solvent liquid ammonia, the sodium enolate from cyclo-hexanone reacts slowly with n-butyl bromide to give a mixture of monobutyl and dibutylcyclohexanone [1]. Allyl bromide reacts faster under these conditions and 2-allylcyclohexanone is obtained in a reasonable yield in addition to some diallyl derivative [2]. Double deprotonation of a symmetrical ketone like acetone by successive treatment with potassium hydride and BuLi·TMEDA gives a species which is far more reactive towards alkyl halides than the initial potassium enolate [3]. β-Diketones undergo a very easy deprotonation at the central carbon atom upon treatment with a variety of basic reagents. The weakly basic $(CO-CH-CO$ $\sim pK$ 9) species are much less reactive in alkylation than enolates derived from mono-ketones like acetone. Interaction between the mono-metallated diketones and a strong base gives rise to the formation of a di-anionic species which exhibit a much higher nucleophilicity than its precursor [4]. Thus, pentane-2,4-dione, for example, is successively converted into $CH_3COCH(Na)COCH_3$ and $NaCH_2COCH(Na)COCH_3$ by reaction with two equivalents of sodamide in liquid ammonia [4]. The subsequent reaction with alkylation reagents RX exclusively takes place at the terminal carbon atom to afford (after aqueous work-up) $RCH_2COCH_2COCH_3$ as shown in the preceding experiment. Dimetallation of β-diketones can be achieved also with two equivalents of LDA in a THF-hexane mixture. The resulting solution is more suitable for regiospecific reactions with other electrophiles (e.g., enolizable aldehydes and ketones) than the ammoniacal solutions. Alkylations with non-activated halides do not proceed readily in THF-hexane mixtures.

Procedure

Pentane-2,4-dione (0.05 mol, 5.0 g, diluted with 20 ml of THF) is added over a few min to a solution of 0.11 mol of LDA (see Exp. 1) in 73 ml of hexane and 60 ml of THF cooled to $-50\,^\circ C$. After the addition, the cooling bath is removed and the temperature allowed to rise to $+10\,^\circ C$. Subsequently, the solution is cooled to $-10\,^\circ C$ and 0.07 mol (8.4 g) of allyl bromide is added in one portion, whereupon the cooling bath is removed. The temperature may rise to above 20 $^\circ C$. The reaction is completed by heating the solution for an additional 15 min at 30–35 $^\circ C$, then the reaction mixture is transferred into a 1-l round-bottom and the greater part of the solvent is removed under reduced pressure (rotary evaporator). A cold $(0\,^\circ C)$ mixture of 16 g of 37% aqueous hydrochloric acid and 120 ml of water is added to the residue with manual swirling. The product is isolated by extraction (four times) with

Et$_2$O, drying the (unwashed) extracts over MgSO$_4$ and distillation. The allylation product, b.p. 75 °C/12 mmHg, n$_D$(22) 1.4763, is isolated in ~ 75% yield.

Literature

1. Verkruijsse HD, Brandsma L (unpublished)
2. Vanderwerf CA, Lemmerman LV Org Synth, Coll. Vol. 3:44
3. Hubbard JS, Harris TM (1980) J Am Chem Soc 102:2110
4. Harris TM, Harris CM (1969) Org Reactions 17:155, John Wiley

5.12 1-Trimethylsilyloxy-1-heptene

$$C_6H_{13}CH{=}O + Me_3SiCl + Et_3N \xrightarrow[\text{reflux}]{\text{DMF}}$$

$$C_5H_{11}CH{=}CHOSiMe_3 + Et_3N \cdot HCl{\downarrow}$$

Apparatus: 500 ml round-bottomed, three-necked flask equipped with a thermometer (dipping in the reaction mixture), a reflux condenser, and a mechanical stirrer; on the top of the condenser a tube filled with KOH pellets is placed.

Procedure

The flask is charged with 0.4 mol (45.6 g) of freshly distilled heptanal, 100 ml of DMF, 0.45 mol (48.6 g) of freshly distilled chlorotrimethylsilane, and 0.5 mol (50.5 g) of triethylamine (dried over KOH). The stirred mixture is heated in an oil bath. The bath temperature is gradually increased in order to maintain a constant reflux. The temperature of the mixture increases over ~ 2 h from ~ 95 to 135 °C. The bath is then removed and the mixture cooled to room temperature. The salt slurry is poured into 500 ml of ice water and the reaction flask rinsed with ice water. After separation of the layers, three extractions with pentane are carried out (Note 1). The combined organic solutions are washed twice with ice water, dried over MgSO$_4$, and subsequently concentrated in vacuo. Careful distillation of the remaining liquid through an efficient column gives, after a small forerun of the aldehyde, the silyl enol ether, b.p. 70 °C/12 mmHg, n$_D$(20) 1.4224, in ~ 75% yield.

Note

1. In view of the water-sensitivity of the product, all operations during the work-up should be carried out quickly.

Metallation-Functionalization Index (Vol. II)

Substrate	Metallic Derivatives*	Metallation Conditions	Functionalization Reagent and Conditions	Page
Chapter 11				
$PhCH_3$	$PhCH_2K$ $PhCH_2Li$	BuLi·t-BuOK, hexane BuLi·TMEDA, hexane	$Et_2O + Me_3SiCl$ $THF + c\text{-}C_6H_{11}Br$ or $BrCH_2CH_2(OEt)_2$	23 30
$PhCH_2CH_3$	$PhCH(K)CH_3$	BuLi·t-BuOK, TMEDA, hexane	$Et_2O + Me_3SiCl$	26
Naphthyl-1-CH_3 Naphthyl-2-CH_3	Naph-1-CH_2K Naph-2-CH_2K	BuLi·t-BuOK, TMEDA, hexane	$THF + Me_3SiCl$	25
$C_6H_4(CH_3)CH_3$ (o,m,p-xylene)	$C_6H_4(CH_3)CH_2Li$	BuLi·TMEDA, hexane	$THF + DMF$, CH_3SSCH_3, $CH_3N=C=S$, Me_3SiCl or CO_2	30
$C_6H_3(CH_3)_2CH_3$ (mesitylene)	$C_6H_3(CH_3)_2CH_2Li$	BuLi·TMEDA	$THF + (H_2C=O)_n$	30
$H_2C=C(CH_3)CH=CH_2$	$H_2C=C(CH_2K)CH=CH_2$	LDA + t-BuOK, THF, hexane	oxirane	43
$PhC(CH_3)=CH_2$	$PhC(CH_2K)=CH_2$	LDA + t-BuOK, THF, hexane	oxirane	44
$H_2C=CH—CH_3$	$H_2C=CH—CH_2K$	BuLi·t-BuOK, THF, hexane	LiBr, then CS_2, then CH_3I, or: LiBr, then $CH_3N=C=S$	33
$H_2C=C(CH_3)CH_3$	$H_2C=C(CH_3)CH_2K$	BuLi·t-BuOK, THF, hexane, or BuLi·t-BuOK, TMEDA, hexane	$(THF +)$ $C_8H_{17}Br$, $c\text{-}C_6H_{11}Br$, $PhCH_2Cl$, $BrCH_2CH(OEt)_2$, oxirane; LiBr, then CO_2 or CS_2, $+CH_3I$; LiBr, then $CH_3N=C=S$	33
$H_2C=C(CH_3)CH_3$ (dimetallation)	$H_2C=C(CH_2K)CH_2K$	2BuLi·t-BuOK, hexane	THF + oxirane	41
$C_6H_4(CH_3)CH_3$ (m-xylene)(dimetallation)	$C_6H_4(CH_2K)CH_2K$	2BuLi·t-BuOK, TMEDA, hexane	$THF + Me_3SiCl$	28

(Continued)

Substrate	Metallic Derivatives*	Metallation Conditions	Functionalization Reagent and Conditions	Page
Chapter II				
$CH_3CH_2CH=CH_2$ $C_4H_9CH_2CH=CH_2$ $Z\text{-}CH_3CH=CHCH_3$	$CH_3CHKCH=CH_2$ $C_4H_9CHKCH=CH_2$ $KCH_2CH=CHCH_3$	BuLi·t-BuOK, THF-hexane, or BuLi·t-BuOK, TMEDA, hexane	Me_3SiCl	36
Limonene	Limonene-K	BuLi·t-BuOK, THF-hexane	Me_3SiCl	36
α-Pinene	α-Pinene-K	BuLi·t-BuOK, hexane	THF + Me_3SiCl	36
1,4-Cyclohexadiene	Cyclohexadiene-K	BuLi·TMEDA, hexane	THF + Me_3SiCl	36
$PhCH_2CH=CH_2$	$PhCHLiCH=CH_2$	BuLi·THF, hexane	Me_3SiCl	38
	$PhCHNaCH=CH_2$	$NaNH_2$, liq. NH_3	C_2H_5Br	39
$CH_3CH=CHCH=CH_2$	$KCH_2CH=CHCH=CH_2$	BuLi·t-BuOK, THF, hexane	$C_6H_{13}Br$	36
Cyclohexene	Cyclohexene-K	BuLi·t-BuOK, THF, hexane	oxirane	40
Indene	Indene-Li	BuLi, THF, hexane	C_4H_9Br, Me_3SiCl	45
Cyclopentadiene	Cp-Li	$LiNH_2$, liq. NH_3	$C_6H_{13}Br$	47
	Cp-Na	NaH, THF	Me_3SiCl	
Chapter III				
$(CH_3S)_2CH_2$	$(CH_3S)_2CHLi$	BuLi, THF, hexane	C_4H_9Br, $H_2C=CHCH_2Br$, $PhCH_2Br$, oxirane, epoxypropane, epoxycyclohexane, $(CH_2O)_n$, DMF, CO_2, CH_3SSCH_3, Me_3SiCl, $CH_3N=C=S$, $ClCONMe_2$, D_2O	59, 61 63, 66 67, 69 70, 72 73
1,3-Dithiane	Dithianyl-Li	BuLi·TMEDA, hexane	—	61
CH_3OCH_2SPh	$LiCH(OCH_3)SPh$	BuLi, THF, hexane	$Cl(CH_2)_3Br$	64
		BuLi, THF, hexane	DMF, Me_3SiCl	72

Substrate	Metalated species	Conditions	Reaction / Reagent	Page
$(CH_3S)_2CHSiMe_3$	$(CH_3S)_2CLiSiMe_3$	BuLi, THF, hexane	$PhCH=O$, $C_2H_5CH=O$, cyclohexanone (Peterson elimin.), cyclohexenone (1,4-add)	74
$(EtS)_2CH_2$	$(EtS)_2CHLi$	BuLi, THF, hexane	D_2O	83
$PhCH(SCH_3)_2$	$PhCLi(SCH_3)_2$	BuLi, THF, hexane	cyclohexenone (1,4-add)	76
$EtSCH_2S(=O)Et$	$EtSCH(Li)S(=O)Et$	BuLi, THF, hexane	$H_2C=CHCOCH_3$(1,4-add)	76
CH_3SCH_3	CH_3SCH_2Li	BuLi·TMEDA, hexane	$THF + PhCH=O$	78
$PhSCH_3$	$PhSCH_2Li$	BuLi·TMEDA, hexane	$THF + Me_3SiCl$	78
$PhSCH_2SiMe_3$	$PhSCHLiSiMe_3$	BuLi, THF, hexane	acetone (Peterson elim.)	79
$PhSCH_3$ (dimetallation)	$o\text{-}Li\text{-}C_6H_4\text{-}CH_2Li$	2BuLi·TMEDA, hexane	$Et_2O + Me_3SiCl$	80
$(CH_3S)_2CH_2$	$(CH_3S)_2CHK$	KNH_2, liq. NH_3	oxirane	81
CH_3SOCH_3	CH_3SOCH_2Na	$NaNH_2$, liq. NH_3	$C_6H_{13}Br$	82
$PhSOCH_3$	$PhSOCH_2Li$	LDA, THF, hexane	C_4H_9Br	85

Chapter IV

Substrate	Metalated species	Conditions	Reaction / Reagent	Page
$C_6H_4(NMe_2)CH_3$ Dimethyl-o-toluidine	$C_6H_4(NMe_2)CH_2Li$	BuLi·TMEDA, hexane	$THF + Me_3SiCl$	93
$C_6H_4(OR)CH_3$ protected o-cresol ($R=CH(CH_3)OEt$)	$C_6H_4(OR)CH_2K$	BuLi·t-BuOK, TMEDA, hexane	$THF + C_4H_9Br$ (C-alkylation)	95
$C_6H_4(OH)CH_3$ o-cresol (dimetallation)	$C_6H_4(OK)CH_2K$	2BuLi·t-BuOK, TMEDA, hexane	$THF + 1\,BuBr$, $2Me_3SiCl$ (bis-silylation)	97
$C_6H_4(C\equiv N)CH_3$ o- or p-tolunitrile	$C_6H_4(CN)CH_2K$	LDA, t-BuOK, THF, hexane or $LiNH_2$, liq. NH_3 LDA, THF, hexane	$C_5H_{11}Br$	99
$C_6H_4(SO_2NMe_2)CH_3$	$C_6H_4(SO_2NMe_2)CH_2Li$	Me₃SiCl	Me_3SiCl	102

Chapter V

Substrate	Metalated species	Conditions	Reaction / Reagent	Page
$H_2C=CHCH_2NMe_2$	$H_2C=CHCH(Li)NMe_2$	BuLi, TMEDA, hexane	$THF + C_5H_{11}Br$	113
$H_2C=CHCH_2NEt_2$	$H_2C=CHCH(K)NEt_2$	BuLi·t-BuOK, THF, hexane	Me_3SiCl	113
$H_2C=CHCH_2Ot\text{-}Bu$	$H_2C=CHCH(K)Ot\text{-}Bu$	BuLi·t-BuOK, THF, hexane	$C_6H_{13}Br$	115

(Continued)

Substrate	Metallic Derivatives*	Metallation Conditions	Functionalization Reagent and Conditions	Page
Chapter V				
$H_2C=CHCH_2SiMe_3$	$H_2C=CHCH(Li)SiMe_3$	BuLi·TMEDA, hexane	$Et_2O + Me_3SiCl$, or THF + cyclohexanone oxirane (R=Ph)	116
$H_2C=CHCH_2SR$ (R=CH$_3$, Ph)	$H_2C=CHCH(Li)SR$	BuLi, THF,		118
$H_2C=C(OCH_3)CH_3$	$H_2C=C(OCH_3)CH_2K$	BuLi·t-BuOK, THF, hexane	$C_8H_{17}Br$	119
$PhCH_2NMe_2$	$PhCH(K)NMe_2$	BuLi·t-BuOK, THF, hexane	CH_3I	120
$PhCH_2SCH_3$	$PhCH(Li)SCH_3$	BuLi, THF, hexane	Me_3SiCl	122
$PhCH_2SiMe_3$	$PhCH(Li)SiMe_3$	BuLi·TMEDA, THF, hexane	Me_3SiCl, cyclohexanone (Peterson elim.)	122
Chapter VI				
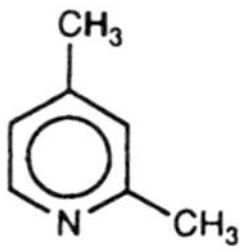 [2-methyl-4-methylpyridine]	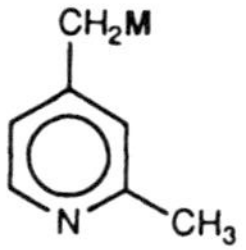 [2-methyl-pyridine-4-CH$_2$M]	NaNH$_2$ or KNH$_2$, liq. NH$_3$ / LDA, THF, hexane / BuLi, THF, then TMEDA	C_3H_7Br, $BrCH_2CH_2CH_2(OEt)_2$, $c\text{-}C_6H_{11}Br$	131, 136
[2-methyl-4-methylpyridine]	[2-methyl-pyridine-4-CH$_2$Li]	BuLi, Et$_2$O, or THF, hexane	Me_3SiCl, CH_3SSCH_3, $CH_3CH=O$	133
$C_5H_4NCH_3$ 3-methylpyridine	$C_5H_4NCH_2Li$	LDA, THF, hexane	CH_3SSCH_3, Me_3SiCl oxirane, I_2 (oxidative coupling), $CH_3CH=O$, $C_2H_5C\equiv N$, then H^+, H_2O / C_4H_9Br	138

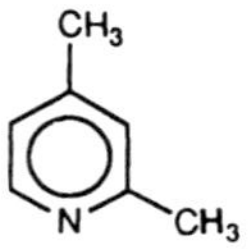

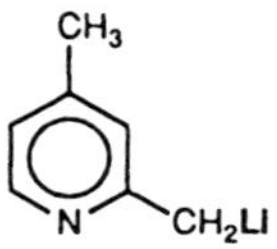

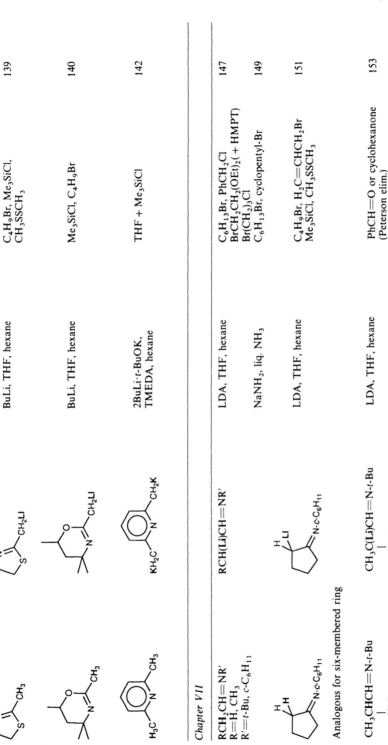

139	BuLi, THF, hexane	C$_4$H$_9$Br, Me$_3$SiCl, CH$_3$SSCH$_3$	
140	BuLi, THF, hexane	Me$_3$SiCl, C$_4$H$_9$Br	
142	2BuLi·t-BuOK, TMEDA, hexane	THF + Me$_3$SiCl	

Chapter VII

147	RCH$_2$CH=NR' → RCH(Li)CH=NR' (R=H, CH$_3$; R'=t-Bu, c-C$_6$H$_{11}$)	LDA, THF, hexane	C$_6$H$_{13}$Br, PhCH$_2$Cl, BrCH$_2$CH$_2$(OEt)$_2$ (+ HMPT), Br(CH$_2$)$_3$Cl
149		NaNH$_2$, liq. NH$_3$	C$_6$H$_{13}$Br, cyclopentyl-Br
151		LDA, THF, hexane	C$_4$H$_9$Br, H$_2$C=CHCH$_2$Br, Me$_3$SiCl, CH$_3$SSCH$_3$

Analogous for six-membered ring

153	CH$_3$CHCH=N-t-Bu / SiMe$_3$ → CH$_3$C(Li)CH=N-t-Bu / SiMe$_3$	LDA, THF, hexane	PhCH=O or cyclohexanone (Peterson elim.)

(*Continued*)

Substrate	Metallic Derivatives*	Metallation Conditions	Functionalization Reagent and Conditions	Page
Chapter VIII				
$RCH_2C\equiv N$ R=H, alkyl, Ph, Et_2N	$RCH(Li, Na)C\equiv N$	$LiNH_2$ or $NaNH_2$ liq. NH_3	$C_6H_{13}Br$, C_4H_9Br, $PhCH_2Br$, c-pentyl-Br, cyclohexanone, CH_3CO-t-Bu, oxirane, epoxypropane, cyclohexeneoxide	161 163 165 167
$RCH_2C\equiv N$ (R also $H_2C=CH$)		LDA, THF, hexane	cyclohexanone, $CH_3C=O$, t-$BuCH=O$, epoxypropane, $PhCH_2Br$, $C_6H_{13}Br$, $Br(CH_2)_3Cl$, C_4H_9I, C_4H_9Br	163, 165 167
$CH_3N=C$	$LiCH_2N=C$	BuLi, THF, hexane (only when R=H or Ph) BuLi, THF, hexane	C_6H_5Br, oxirane, cyclohexanone	162, 168
Chapter IX				
H_2CCl_2 $RCHCl_2$ $R=CH_3, C_8H_{17}$	$LiCHCl_2$ $RCLiCl_2$	BuLi, THF, hexane BuLi:TMEDA, THF, hexane	$PhCH=O, C_8H_{17}Br$ (+HMPT) t-$C_4H_9CH=O, D_2O$	176 177
H_2CBr_2 $HCCl_3$ $HCBr_3$	$LiCHBr_2$ $LiCCl_3$ $LiCBr_3$	LDA, THF, hexane LDA, THF, hexane LDA, THF, hexane	$HMPT + C_4H_9Br; Me_3SiCl$ $HMPT + C_5H_{11}Br$ $HMPT + C_4H_9Br$	178 179 180
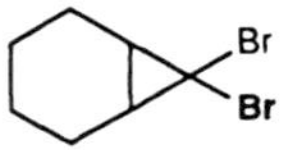 (cyclopropane: Br, Br)	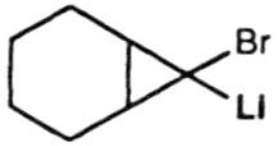 (cyclopropane: Br, Li)	BuLi, THF, Hexane + LiBr	Me_3SiCl, DMF	181
$BrCH_2COOEt$	$BrCH(Li)COOEt$	LDA, THF, hexane	cyclohexanone	183

Chapter X

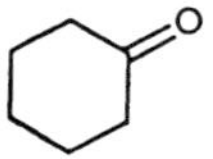

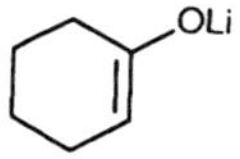

Substrate	Enolate	Conditions	Electrophile	Ref.
cyclohexanone	(1-cyclohexenyl OLi)	LDA, THF, hexane	Me₃SiCl	190
2-methylcyclohexanone	(methyl cyclohexenyl OLi)	LDA, THF, hexane	Me₃SiCl	190
PhCOCH₃	PhC(OLi)=CH₂	LDA, THF, hexane	Me₃SiCl	190
C₉H₁₉COCH₃	C₉H₁₉C(OLi)=CH₂	LDA, THF, hexane	Me₃SiCl	190
RCH=CHCOCH₃	RCH=CHC(OLi)=CH₂	LDA, THF, hexane	Me₃SiCl	190
β-ionone				
C₆H₁₃C≡CCOCH₃	C₆H₁₃C≡CC(OLi)=CH₂	LDA, THF, hexane	Me₃SiCl	190
C₅H₁₁CH=CHOSiMe₃	C₅H₁₁CH=CHOLi	BuLi, Et₂O, hexane	(CH₃CO)₂O	192
RCH₂COOR'	RCH=C(OR')OLi	LDA, THF, hexane		193
R=H, R'=C₂H₅			PhCH=O, cyclohexanone	
R=CH₃, R'=t-C₄H₉			PhCH=O	
R=C₂H₅, R'=CH₃			I₂,Me₃SiCl, H₂C=CHCH₂Br	
CH₃CH=CHCOOCH₃	H₂C=CHCH=C(OCH₃)OLi	LDA, THF, HMPT, hexane	C₄H₉I	199
CH₃COO-t-C₄H₉	H₂C=C(O-t-C₄H₉)OLi	LiNH₂, liq. NH₃	C₄H₉Br, PhCH₂Cl, Cl(CH₂)₃Br	196
CH₃CONMe₂	H₂C=C(NMe₂)OLi	LiNH₂, liq. NH₃	(CH₃)₂CHBr, C₈H₁₇Br, Cl(CH₂)₃Br	196
(pyrrolidinyl cyclopentanone)	(cyclopentenyl-N-CH₃ OLi)	LDA, THF, hexane	H₂C=CHCH₂Br	197
(γ-butyrolactone)	(dihydrofuran OLi)	LDA, THF, HMPT, hexane	H₂C=CHCH₂Br	197

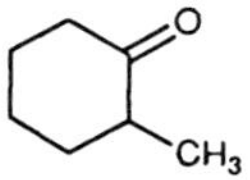

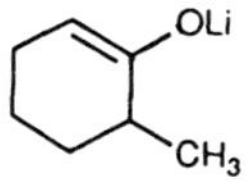

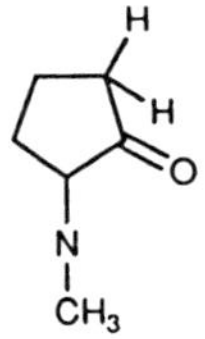

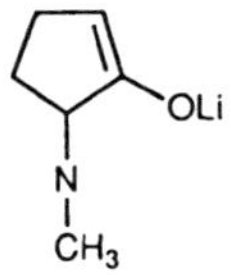

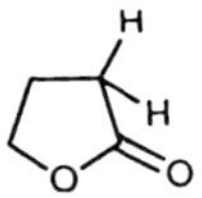

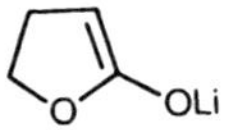

(Continued)

Substrate	Metallic Derivatives*	Metallation Conditions	Functionalization Reagent and Conditions	Page
Chapter X				
$CH_3CH_2C(=S)OCH_3$	$CH_3CH=C(OCH_3)SLi$	LDA, THF, hexane	$H_2C=CHCH_2Br$ (3,3-sigm. rearr.)	200
$CH_3(C=O)SEt$	$H_2C=C(SEt)OLi$	LDA, THF, hexane	$t-C_4H_9CH=O$	200
$CH_3(C=S)SEt$	$H_2C=C(SEt)SLi$	LDA, THF, hexane	$t-C_4H_9CH=O$ (elim. of water)	200
$C_3H_7(C=S)S'CH_3$	$C_2H_5CH=C(SCH_3)SLi$	LDA, THF, hexane	Me_3SiCl	200
RCH_2CSSR'	$RCH=C(SR')SM$	MNH_2; liq. NH_3 $M=Li$, Na	$R''Br$ S-alkylation	202
$C_4H_9CH_2COOH$	$C_4H_9CH=C(OLi)ONa$	NaH + LDA, THF, hexane	$H_2C=CHCH_2Br$	203
$CH_3COCH_2COCH_3$	$NaCH_2C(ONa)=CHCOCH_3$	$2NaNH_2$, liq. NH_3	C_3H_7Br (terminal alkylation)	204
$CH_3COCH_2COCH_3$	$LiCH_2C(OLi)=CHCOCH_3$	2LDA, THF, hexane	$H_2C=CHCH_2Br$	205

* The abstracted atom in the substrate and the introduced metal atom in the metallic derivative are printed in bold (with the exception of the formulas in Chapter X).

Syntheses of Reagents and Starting Compounds (Vols. I and II)

(Roman letters I and II refer to volume I and II respectively)

(Z-)BrCH=CHOEt, I-72
PhC(Br)=CH$_2$, I-73
(CH$_3$)$_2$C=CHBr, I-66
HC≡CCH=CH$_2$, I-67
1-Methylcyclopropene, I-67
CH$_3$C≡CC$_6$H$_{13}$, I-65
1-Chlorocyclohexene, I-65
1-Chlorocycloheptene, I-65
5-Chloro-2,3-dihydropyran, I-106
H$_2$C=CHSCH$_3$, I-107
H$_2$C=CHSC$_2$H$_5$, I-107
H$_2$C=CHSPh, I-107
PhCH$_2$CH=CH$_2$, II-51
(Z-)CH$_3$SCH=CHSCH$_3$, I-108
2H-Thiopyran, I-109
[(CH$_3$)$_2$N]$_2$C=CHCl, I-110
(CH$_3$)$_2$NCH$_2$C=CC(CH$_3$)$_2$OCH$_3$, I-111
3-Bromothiophene, I-176
2-Bromothiophene, I-177*
3-Bromoquinoline, I-177
2H-Thieno[2,3-b]thiopyran, I-178
2-Methylthienothiopnene, I-178
Methyl dithiocarbamate (NH$_2$CSSCH$_3$), I-179
2-Methylthiothiazole, I-180
4-Methyl-2-methylthiothiazole, I-180
Thiazole, I-181
4-Methylthiazole, I-181
Selenophene, I-182
Tellurophene, I-182

Methyl thiocyanate (CH$_3$SC≡N), I-183
Sulfur dichloride (SCl$_2$), I-184
PhOCH(CH$_3$)OC$_2$H$_5$, I-199
o-Cl-C$_6$H$_4$-O-CH$_2$OCH$_3$, I-199
PhSO$_2$N(CH$_3$)$_2$, I-209
PhS-t-Bu, I-200
1,4-Cyclohexadiene, II-50**
H$_2$C(SCH$_3$)$_2$, II-86
H$_2$C(SC$_2$H$_5$)$_2$, II-86
H$_2$C(SPh)$_2$, II-87
C$_2$H$_5$SCH$_2$SOC$_2$H$_5$, II-87
PhCH(SCH$_3$)$_2$, II-88
CH$_5$SOPh, II-87
p-CH$_3$-C$_6$H$_4$SO$_2$N(CH$_3$)$_2$, II-103
H$_2$C=CHCH$_2$SiMe$_3$, II-126
H$_2$C=CHCH$_2$O-t-Bu, II-124
H$_2$C=CHCH$_2$N(CH$_3$)$_2$, II-123
H$_2$C=CHCH$_2$N(C$_2$H$_5$)$_2$, II-123
H$_2$C=CHCH$_2$SCH$_3$, II-125
CH$_3$SCH$_2$Ph, II-125
H$_2$C=CHCH$_2$SPh, II-125
CH$_3$CH=N-t-Bu, II-155
C$_2$H$_5$CH=N-t-Bu, II-155
CH$_3$CH=N-c-C$_6$H$_{11}$, II-155
(CH$_2$)$_4$C=N-c-C$_6$H$_{11}$, II-155
(CH$_2$)$_5$C=N-c-C$_6$H$_{11}$, II-155
CH$_3$CH(C≡N)OCH(CH$_3$)OC$_2$H$_5$, II-170
CH$_3$CH(OH)C≡N, II-170
C$_5$H$_{11}$CH=CH—OSiMe$_3$, II-207

---

* For an improved method see Keegstra M.A., Brandsma L. (1988) Synthesis, 890
** See also Brandsma L., van Soolingen J., Andringa H. (1990) Synth. Comm., 2165

Complementary Subject Index (Vols. I and II)

Activation

1. Activation of Mg in Grignard preparations by addition of $HgCl_2$. Magnesium turnings activated in this manner react at temperatures in the region of $0\,°C$ with halides, e.g.

   H_2C=$CHCH_2Br$ (preparation of H_2C=$CHCH_2MgBr$, Vol. II, p. 126).

2. Mechanical activation of lithium by addition of pieces of broken glass in the preparation of 1-lithiocycloalkenes from chlorocycloalkenes and lithium chips, Vol. I, p. 50.

Addition in Reversed Sense

1. In the reaction of indenyllithium with butyl bromide, to avoid introduction of a second butyl group (Vol. II, p. 45).
2. In reactions with CH_3SSCH_3, to avoid introduction of a second CH_3S group. Example: Vol. II, p. 133.
3. In reactions with iodine to avoid oxidative 'dimerization'. Example: Vol. II, p. 193 (compare p. 135!).
4. In reactions with $ClCOOCH_3$ to avoid further reaction of the organo alkali compound with the ester $RCOOCH_3$. Example: Vol. I, p. 145.
5. In reactions with gaseous CO_2 to avoid reaction of RLi with the carboxylate RCOOLi. Example: Vol. I, p. 62, Vol. II, p. 30, 33, 69.
6. In reactions with $ClCONMe_2$ to avoid reaction of RLi with $RCONMe_2$. Examples: Vol. I, p. 146; Vol. II, p. 73.

Azeotropic Removal of Water

Applied in the preparation of ketimines from cycloalkanones and cyclohexylamine. Examples: Vol. II, p. 155.

Drying (see also Vol. I, p. 6–8)

Potassium carbonate is generally used for drying amines, especially when the product is very acid-sensitive, e.g. RCH=$CHNR'_2$ which may undergo Z to E conversion under the influence of magnesium sulfate. Example: Vol. II, p. 113.

Dissolving Alkali Metal Reductions in Liquid Ammonia

Examples: Vol. I, p. 181 and Vol. II, p. 50.

Solvents (see also Vol. I, p. 6)

1. Use of HMPT as co-solvent: several examples have been described in Vol. I and II, e.g. the alkylation of o-potassio fluorobenzene at $< -80\,°C$, Vol. I,

p. 206 and the alkylation of $LiCBr_3$ at temperatures between -100 and $-110\,°C$, Vol. II, p. 178–180.

2. Chloroform as extraction solvent.

Applied to isolate certain sulfoxides from aqueous solutions, when partitioning between water and Et_2O is unsatisfactory. Examples: Vol. II, p. 83, 87.

3. Dichloromethane as extraction solvent.

Applied in the work-up of reaction mixtures resulting from syntheses in liquid ammonia starting with the preparation of potassium amide: very small particles of potassium often remain unconverted since they are covered by oxide. If, after addition of water, Et_2O is used for the extraction, this unconverted potassium may give rise to fire hazards. Examples: Vol. I, p. 55; Vol. II, p. 82.

4. Paraffin oil as additive in destillative work-up.

Used as a heat-conducting liquid when the product has to be distilled off from solid material. Example: Vol. I, p. 102.

Also applied to minimize the risk of explosive decomposition in the last stage of the distillation of thermolabile compounds. Example: Vol. I, p. 105

5. Liquid NH_3 as a solvent (see also Vol. I, p. 6, Vol. II, p. 5)

Several examples described in Vol. I and II. For reactions carried out at temperatures below the b.p. $(-33\,°C)$ of ammonia see Vol. I, p. 181; Vol. II, p. 82, 108, 196.

6. High-boiling petroleum ether as extraction solvent.

Used in the isolation of compounds with a b.p. < 100 to $110\,°C/760\,mmHg$. Examples: Vol. I, p. 48, Vol. II, p. 50.

Work-up (see also under solvents)

1. 'Dry' work-up: applied when the product is water-sensitive, e.g. ketene-acetals $>C=C(OSiMe_3)(OR)$, Vol. II, p. 193; see also Vol. II, p. 168.

2. Addition of glacial acetic acid to liberate alcohols from ROLi.

Applied when the alcohol can undergo base-catalyzed cyclization during addition of water. Example Vol. II, p. 169.

3. Addition of t-butyl alcohol to liberate compounds from their metallic derivatives. Applied when the product is water-sensitive. Example Vol. I, p. 102.

4. Addition of a limited amount of water dissolved in THF to liberate alcohols from ROLi. Applied when the product has a very good water-solubility. Examples: Vol. II, p. 134, 168.

5. Addition of solid NH_4Cl to liberate alcohols from ROM prepared in liquid ammonia.

Applied when the product has a good water-solubility, making extraction laborious. Example: preparation of $HO(CH_2)_3C\equiv N$, Vol. II, p. 168.

6. Removal of the greater part of the solvents in vacuo prior to aqueous work-up.

Applicable in the cases of rather volatile products (b.p. up to $\sim 60\,°C/15\,mmHg$) which after the removal of the solvents and the addition of water can be extracted with limited amounts of volatile solvents. Example: Vol. II, p. 151.

Also applied after reactions with excess of Me_3SiCl, when the presence of this compound during the aqueous work-up is undesired, see Vol. II, p. 10.

7. Aqueous work-up with a $KC\equiv N$ solution.

Applied to remove copper salts used in the synthesis. Examples: Vol. I, p. 148, 150.

8. Aqueous work-up with KOH solution.
   Applied to prevent liberation of thiol from RSLi during the addition of water. Example: Vol. I, p. 150.

9. Removal of THF by repeated washing with water.
   Applied in the preparation of H_2C=$CHSiMe_3$ which cannot be separated from THF by distillation: Vol. I, p. 48.

Some Typical Synthetic Procedures (see also under 'Addition in a reversed sense')

1. Deuteration: Vol. I, p. 172, Vol. II, p. 83.

2. a. Protection of OH-groups with H_2C=$CHOC_2H_5$: Vol. II, p. 95.
   b. Deprotection: Vol. II, p. 95.

3. pH-controlled conversion of imines into the aldehydes or ketones. Vol. II, p. 152.

4. Formylation with DMF, procedures for the acid hydrolysis of the adduct formed from RLi and HC(=O)NMe$_2$. Examples: Vol. I, p. 51–54, 142 and Vol. II, p. 67, 72.

5. Birch-reduction of benzene to 1,4-cyclohexadiene: Vol. II, p. 50.

6. Chlorination with Cl_3CCCl_3: Vol. I, p. 168.

7. Use of CH_3SC≡N for introduction of a CH_3S group (instead of CH_3SSCH_3): Vol. I, p. 169.

8. Replacement of Li by MgBr in order to prevent introduction of a second iodine atom: Vol. I, p. 157. Also applied to introduce a O-t-C_4H_9 group by reaction with PhC(=O)OO-t-C_4H_9: Vol. I, p. 158.

9. EtLi·LiBr in Et$_2$O as metallating agent. Used when the product of the synthesis is volatile so that the presence of hexane from the commercially available BuLi is undesired: Vol. I, p. 121. Also applied as a mild reagent in the lithiation of tellurophene, which compound react less selectively with the salt-free BuLi in THF or Et$_2$O-hexane mixtures: Vol. I, p. 143.

10. Potassium diisopropylamide as metallating agent.
    A 1:1 molar mixture of t-BuOK and LDA effects a more complete metallation of isoprene (Vol. II, p. 43) and α-methyl styrene (Vol. II, p. 44) than does LDA alone.

11. Liq. NH$_3$ as a solvent, see under Solvents.

Cooling Technique

For a detailed description of the cooling technique applied in reactions at very low temperatures see Vol. II, p. 175.

Purification and Storage of Some Reagents and Solvents
(see also Vol. I, Chap. I)

Acetaldehyde
This aldehyde polymerizes very easily giving the cyclic trimer 'paraldehyde'. For this reason it is absolutely necessary to use the *freshly distilled* compound in reactions (b.p. 21 °C/760 mmHg). Since traces of acid adhering to the glass may catalyze the trimerization, the condenser and receiver should be rinsed first with a dilute solution of triethylamine or diethylamine in acetone and subsequently blown dry. Monomeric acetaldehyde may also be obtained by adding 96% H_2SO_4 (1 ml) with manual swirling to 'paraldehyde' (100 to 300 ml) and subsequent distillation through a 40-cm Vigreus column.

Aliphatic Aldehydes
Since trimerization and oxidation occur during storage, it is always necessary to distill the compounds prior to use: the lower homologues at atmospheric pressure, hexanal and higher homologues at 10 to 20 mmHg.

Acetonitrile (and higher homologues)
The *only* effective drying agent for acetonitrile is *phosphorus pentoxide*. It may also be used for the homologues. The following procedure is recommended: the nitrile (200 to 500 ml) is vigorously shaken with $P_2O_5(\sim 5\,g)$ until the suspended material has clustered to a slurry. The liquid is decanted and shaken with a second portion of $\sim 3\,g$ of P_2O_5. The procedure is repeated until the P_2O_5 remains in suspension. The nitrile is then distilled off from the P_2O_5 in a vacuum of 10 to 30 mmHg and collected in a receiver cooled below $-40\,°C$ (a tube filled with KOH pellets should be placed between the receiver and the water aspirator). The contents of the receiver may be redistilled at atmospheric pressure. The solvent should be stored at room temperature in a bottle closed with a well-greased glass stopper.

Alkenes (see olefinic compounds)

Amines
A very simple way to investigate the water content of amines is to shake them vigorously with KOH pellets (30 g of pellets for each 100 g amine). If a turbidity or suspension appears, appreciable amounts of water are present and a drying procedure as described in Vol. I, p. 8 is necessary. The amines should be stored on KOH pellets ($\sim 50\,g$ for each 200 to 300 ml).

Benzaldehyde
During storage some benzoic acid may be formed by autoxidation, particularly when the bottle has been used frequently. The acid may be removed by shaking the

aldehyde (100 to 300 ml, diluted with 20 to 50 ml of Et_2O) with a 2% aqueous solution of KOH. The upper layer is dried over $MgSO_4$ and subsequently distilled in vacuo. The purified aldehyde should be stored under nitrogen or argon in a *stoppered* bottle.

n-Butyllithium in Hexane
The commercially available solution in hexane (mostly 1.6 M solution) should always be stored *at room temperature* in well-closed flasks (see Vol. I, p. 10). Screw-capped bottles (with a content of 100, 500 or 1000 ml) should *never* be stored in the *refrigerator* since air may leak inside.

Chlorotrimethylsilane
During storage at variable temperatures moisture may enter the usual screw-capped bottles giving rise to the formation of hydrogen chloride and $Me_3SiOSiMe_3$. A good quality can be generated by distilling the compound at atmospheric pressure from a small amount (5 to 10%) of a non-volatile amine, e.g. N,N-diethylaniline. The distilled sample should be stored at *room temperature* in a bottle with a *well-greased stopper*.

Carbon Tetrachloride and Deuteriochloroform
CCl_4 and $DCCl_3$ are the usual solvents for (routine) NMR measurements. During storage small amounts of HCl and DCl may be formed, which in some cases may catalyze certain conversions, e.g. $Z \leftrightarrow E$ isomers of alkenyl ethers $RCH{=}CHOR'$ and alkenylamines $RCH{=}CHNR'_2$. Purification, if judged to be relevant, may be carried out by successively shaking the solvents with a few ml of a solution of aqueous ammonia and water and drying over anhydrous sodium sulfate.

Lithium Bromide (anhydrous)
The commercially available anhydrous salt is seldom perfectly 'dry'. The water can be removed as follows. The salt (maximally 50 g) is put in a 1-l round bottom, which subsequently is evacuated ($<$1 mmHg) and heated for 30–60 min in an oil bath at 150 °C. Heating with an infrared lamp may also be carried out. During the period of heating the flask is occasionally swirled manually. The salt should be stored in a bottle closed with a *rubber stopper*.

HMPT $[(CH_3)_2N]_3P{=}O$
Since a negligible water content of this hygroscopic liquid cannot be guaranteed after storage in the usual screw-capped bottle, the solvent has to be purified, and subsequently stored (without molecular sieves!) at room temperature in a bottle with a well-greased stopper.

Purification can be carried out as described in Vol. I, p. 7, but also in the following manner. The solvent (1 l) is placed in a 3-l round-bottomed, three-necked flask, equipped with a gas-inlet tube, a mechanical stiirer and a gas outlet (stopper perforated by a hole of at least 7 mm). A vigorous stream of nitrogen is passed for a few min through the stirred liquid, then a very concentrated solution of lithium in liquid ammonia (e.g. 4 g per 150 ml) is cautiously added in small portions, with stirring, while continuing the introduction of N_2 at a moderate rate (during this operation the outlet is removed and the solution of Li poured through the neck). The addition is stopped when a uniformly blue solution (the blue colour should persist for at least ten min) has formed. The flask is then equipped for a vacuum distilation: 20-cm Vigreux column, condenser and receiver. After the dissolved ammonia has been removed in a water-pump vacuum, the HMPT is distilled at $\leqslant 1$ mmHg (b.p. $< 70\,°C$). A small gelatinous residue should be left behind. After one redistillation pure HMPT is obtained.

Olefinic Compounds
Most olefins form high-boiling products during storage in normal screw-capped bottles due to contact with oxygen and should therefore be distilled prior to use in a synthesis.

Polymeric Formaldehyde ('paraform')
Only the dry-looking fine powder should be used for reactions with organolithium compounds. If desired the powder may be subjected to a drying procedure involving heating for ~ 1 h in a vacuum of 10 to 20 mmHg using a rotary evaporator (bath temperature between 50 and 60 $°C$). Under these conditions some depolymerization may occur.

Potassium tert-butoxide
The self-made reagent (from K and t-BuOH) is the 1:1 complex t-BuOK·HOt-Bu. The uncomplexed base is difficult to make from this complex and can better be purchased. A good quality can be maintained *only* when the reagent is stored (preferably under inert gas) in a flask with a *rubber* stopper.

L. Brandsma, H. D. Verkruijsse, University of Utrecht

Preparative Polar Organometallic Chemistry

Volume 1

Foreword by P. v. Ragué Schleyer

1987. XIV, 240 pp. Softcover DM 78,– ISBN 3-540-16916-4

Contents: Introduction to Vol. 1 and Instructions for the Use of this Book and its Indexes. – Organometallic Reagents, Solvents and Laboratory Equipment. – Reactivity of Polar Organometallic Intermediates. – Metallated Olefinic and Allenic Hydrocarbons. – Metallation of Hetero-Substituted Unsaturated Systems. – Metallated Hetero-Aromatic Compounds. – Metallated Aromatic Compounds. – Appendix. – References.

Preparative Polar Organometallic Chemistry is a selection of proven laboratory procedures for the synthesis and functionalization of organo-alkali and Grignard reagents. All procedures are worked out and checked in the authors' own laboratory. The book is the first volume of a series of laboratory manuals for students in chemistry at higher educational levels and for researchers in organic and inorganic chemistry.

In this first volume the chemistry of polar organometallic intermediates derived from sp^2-compounds is presented. Each chapter starts with a general introduction on the synthetic potential of these reagents. Instructions on safe handling and disposal complement the presentation.

R. Scheffold, University of Bern (Ed.)

Modern Synthetic Methods 1986

Volume 4

1986. VIII, 356 pp. Softcover DM 74,–
ISBN 3-540-16526-6
(Vols. 1–3 were published by Verlag
Sauerländer, Aarau, Switzerland)

Contents: Sound and Light in Synthesis: *K. S. Suslick:* Ultrasound in Synthesis. – *K. Schaffner, M. Demuth:* Photochemically Generated Building Blocks. – *I.-M. Demuth:* Photochemically Generated Building Blocks II. – Synthesis of Enantiomerically Pure Compounds with C,C Bond Formation: *D. Seebach, R. Imwinkelried, T. Weber:* EPC Syntheses with C,C Bond Formation via Actals and Enamines. *G. Helmchen, R. Karge, J. Weetman:* Asymmetric Diels-Alder Reactions with Chiral Enoates as Dienophiles. *H. C. Brown, P. K. Jadhav, B. Singram:* Enantiomerically Pure Compounds via Chiral Organobones.

This volume is the conference documentation of the fourth Seminar on **Modern Synthetic Methods.**

R. Scheffold, University of Bern (Ed.)

Modern Synthetic Methods 1989

Volume 5

1989. VII, 304 pp. 24 figs. 8 tabs. Softcover DM 78,– ISBN 3-540-51060-5

Contents: *D. H. G. Crout, M. Christen:* Biotransformations in Organic Synthesis. – *R. Noyori, M. Kitamura:* Enantioselective Catalysis with Metal Complexes, an Overview. – *A. Pfaltz:* Enantioselective Catalysis with Chiral Co- and Cu-Complexes. – *J. M. Thomas, C. R. Theocharis:* Clays, Zeolites and Other Microporous Solids for Organic Synthesis.

This paperback is the conference documentation of the fifth Seminar on **Modern Synthetic Methods.** It is the aim of these triennial Interlaken Seminars to provide a proper, concise and ready-for-use access to important and rapidly developing areas of synthetic organic chemistry.

Springer-Verlag
Berlin
Heidelberg
New York
London
Paris
Tokyo
Hong Kong
Barcelona

Printed in the United States
65720LVS00003B/245